SOIL IN THE ENVIRONMENT

Crucible of Terrestrial Life

SOIL IN THE ENVIRONMENT
Crucible of Terrestrial Life

Daniel Hillel

Goddard Institute for Space Studies
Columbia University
New York, New York

ELSEVIER

Amsterdam • Boston • Heidelberg • London • New York • Oxford
Paris • San Diego • San Francisco • Singapore • Sydney • Tokyo
Academic Press is an imprint of Elsevier

Academic Press is an imprint of Elsevier
30 Corporate Drive, Suite 400, Burlington, MA 01803, USA
525 B Street, Suite 1900, San Diego, California 92101-4495, USA
84 Theobald's Road, London WC1X 8RR, UK

This book is printed on acid-free paper. ∞

Library of Congress Cataloging-in-Publication Data

Hillel, Daniel.
    Soil in the environment/Daniel Hillel.
        p. cm.
    Includes bibliographical references and index.
    ISBN-13: 978-0-12-348536-6 (hardback : alk. paper)   1. Soils—Environmental aspects. I. Title.
    S596.H55 2007
    631.4—dc22   2007037235

British Library Cataloguing-in-Publication Data
A catalogue record for this book is available from the British Library.

ISBN: 978-0-12-348536-6

For information on all Academic Press publications
visit our Web site at www.books.elsevier.com

Printed in China
08 09 10 9 8 7 6 5 4 3 2 1

I dedicate this book to the indelible memory
of my first teacher and mentor in pedology,
"Doc" Jacob Joffe,
profound scientist, passionate environmentalist,
and compassionate humanitarian.
He observed, felt, tasted and listened to the soil,
inhaled its vapors, sensed its life-giving quality,
and conveyed his insights to his students
with contagious and irresistible enthusiasm.
In the classroom, the laboratory, and the field,
he was, as is surname implies, a connoisseur
and aficionado of nature's beauty and vitality.

CONTENTS

PREFACE

We shall not cease from exploring
and the end of all our exploring
will be to arrive where we started
and know the place for the first time.
T.S. Eliot (1888–1965)

The study of the soil has long been regarded as an essential component of agronomic science. Yet it is much more than that, for the soil is—in a larger sense and context—the earth's dynamic processor and recycler of vital materials and energy, the seminal crucible within which life in the land domain is generated and sustained. Hence soil science should, indeed must, be included in the curricula of all students of the life sciences and the environmental sciences, including biology, ecology, geography, geology, climatology, hydrology, as well as civil and environmental engineering, and even the humanities.

The soil of each location on earth is a complex and dynamic system, driven by natural processes and affected by human activities. The components of the soil are a solid phase consisting of particles of various minerals, sizes, and assemblages in association with organic matter; an air phase of varying volume and gaseous composition; and a liquid phase of varying volume, solute composition, and concentration. The soil hosts an enormously diverse community of plants and animals—microscopic, mesoscopic, and macroscopic. The entire seething and intricate complex is the very foundation of our existence on earth, yet it is labile and vulnerable. The future of our civilization depends critically on our ability to understand the nature and workings of the soil and to manage it judiciously so as to avoid its degradation and to maintain its functioning and biological productivity.

In this book, we endeavor to reveal the vital role that the soil fulfills in nature and in human life. We shall describe the essential features and functions of different soil types, consider what makes them suitable for various purposes, explain how soils are used and often abused, and focus on ways to manage soils efficiently while conserving or even enhancing their productivity and avoiding damage to interlinked domains of the open environment.

A textbook on so vital a subject ought by right to convey the sense of wonderment and excitement that impels the scientist's quest to comprehend the workings of nature, and hence should give some pleasure in the reading. It is my hope that this book might indeed be read, not merely consulted, that its readers might find in it a few insights as well as facts, and that it will deepen their understanding as well as broaden their knowledge.

ACKNOWLEDGEMENTS

*Speak to the earth
and it will teach you.*
Job 12:8

My personal fascination with the topic of this book began in my early child-hood, spent in a pioneering settlement in the Jezreel Valley of Israel. It was in this historical frontier between wilderness and habitable land that I was first captivated by the open environment with its vivid contrasts: barren slopes and lush forests, searing droughts and drenching rains, soft sands and soggy swamps, parched plains and fertile fields.

So it was that when the time came for me, as a precocious graduate student, to focus my studies, I took up soil science at Rutgers University. My teachers there were outstanding leaders of this multi-faceted science in their generation: pedologist Jacob Joffe and microbiologist Selman Waksman (both had been students of the illustrious Jacob Lipman, who had established the journal Soil Science in 1916 and organized the first International Congress of Soil Science in 1927), soil chemist Firman Bear, plant physiologist John Shive, climatologist Erwin Biel, and soil physicist George Blake.

Armed with a master's degree from that august university, I first ventured to apply the science in the arid lands of the Middle East. There I had the good fortune to receive the guidance of the visionary soil conservationist Walter Clay Lowdermilk. My doctoral advisors were the innovative Israeli pedolo-gists Avraham Reifenberg and Shlomo Ravikovitch, hydrologist Jacob Rubin, botanist Michael Zohari, plant physiologist Michael Evenari, and ecologist Naftali Tadmor.

In the ensuing five decades, I have had the opportunity to study and inter-act with several leading American soil scientists: Hans Jenny and Geoffrey Bodman at the University of California; L.A. Richards and Wilford Gardner at the U.S. Salinity Laboratory; Francis Hole, Champ Tanner, and Edward Miller at the University of Wisconsin; and Cornelius van Bavel at Texas A&M University; as well as plant physiologists Leon Bernstein and Emanuel Epstein in California. I have also had professional contacts with, and learned much from, outstanding European scientists, among them Dutch physicochemist Gerry Bolt, ecologist-modeler Kees de Wit, and Wim Sombroek of the U.N Food and Agriculture Organization; Howard Penman and Charles Pereira of England; Marcel De Boodt of Belgium; and Walter Sohne of Germany. Others with whom I had cooperated closely were David Hopper and Shawki Barghouti of the World Bank, Tosio Cho and Tomohisa Yano of Japan, Yihya Barrada and Muhammad Mursi of Egypt, Shihada Dajani of Palestine, and Cecil Hourani in Jordan.

All of them and many others by now too numerous to list, challenged and inspired me by their insights and devotion to the science and its application in practice. With the passage of the decades, alas, many of them have passed on, yet their exemplary work, wisdom, and generosity of spirit, live on.

Over the many years, I have had the privilege of teaching hundreds of undergraduate students and advising scores of graduate students in several leading universities, and of working in research and development in over thirty countries around the world. Through these experiences, I have found ample confirmation of the old Mishnaic adage: "Much have I learned from my teachers, yet more from my colleagues, but most of all from my students." Now, as I approach the culmination of my own career, I express my humble gratitude and pay tribute to all the earlier and the present-day practitioners of soil science in its many applications. It is a noble quest, dedicated to understanding the dynamics of the terrestrial environment, to sustaining the intricate-interdependent community of life on earth, and to ensuring the harmonious integration of humanity within this community on our one and only home planet.

1. SOIL AS A LIVING BODY: ITS DYNAMIC ROLE IN THE ENVIRONMENT

To see a world in a grain of sand
and a heaven in a wild flower...
William Blake (1757–1827)
Auguries of Innocence

CRUCIBLE OF TERRESTRIAL LIFE

A romantic poet gazing through his window at a green field outside might view it as a place of idyllic serenity. Not so the environmental scientist, who discerns not rest but unceasing turmoil, a seething foundry in which matter and energy are in constant flux. Radiant energy from the sun streams into the field, and, as it cascades through the atmosphere-plant-soil-subsoil-bedrock continuum, it generates a complex sequence of processes: Heat is exchanged; rainwater percolates through the intricate pores of the soil; plant roots suck up some of that water (along with soluble nutrients); the plants' stems transmit it to the leaves; and the leaves transpire the water back into the atmosphere. The leaves also absorb carbon dioxide from the air and synthesize it with soil-derived water to form the primary compounds of life. Oxygen emitted by the leaves as a consequence of photosynthesis makes the air breathable for animals, which consume and fertilize the plants in return. Organisms in the soil recycle the residues of both plants and animals, releasing nutrients for the perpetual regeneration of life. The remainder of the water—that which is not taken up by plants nor evaporated directly from the soil—drains into groundwater aquifers, which eventually discharge into streams, lakes, and the sea.

The crucible of this foundry is the soil—a mix of mineral particles, organic matter, gases, and nutrients—that, when infused with water, constitutes a fertile substrate for the initiation and maintenance of terrestrial life. The soil is a self-regulating biological factory, utilizing its own materials, water, and energy from the sun. It determines whether and how water from rainfall or snowfall reaching the land's surface will flow downslope as surface runoff or seep into underlying porous rock strata. The soil, with its capacity to absorb, store, and transmit water and solutes, regulates all these phenomena, while serving as a bank for the moisture and nutrient requirements of growing plants and numerous other organisms. Without soil as a buffer, most of the rain falling over the continents would run off directly, producing violent floods rather than sustained river flow. Without soil, the majority of terrestrial creatures—including humans—could not exist. Even creatures living in an aquatic medium depend on nutrients released from the soil into streams, lakes, and seas. Most of our supplies of food and fiber come from the soil, either directly as products obtained from plants grown in the soil or via animals that consume such plant products.

The soil is the terrestrial domain's main recycling medium, a "living filter" wherein pathogens and toxins that might otherwise foul our environment and endanger its myriad forms of life are rendered harmless and transmuted into nutrients. Since time immemorial, humans and other animals have been dying of all manner of diseases and then buried in the soil, yet the soil acts as a disinfectant. The term *antibiotic* was in fact coined by soil scientists who extracted therapeutic products from organisms living in the soil. Ion exchange, a useful process of purifying water, was also discovered by soil scientists studying the passage of electrolytic solutions through beds of clay.

SOIL AS A CENTRAL LINK IN THE BIOSPHERE

Just what do we mean by "soil"? A precise definition is elusive, for what is commonly called "soil" is anything but a homogeneous entity. It is, rather, an exceedingly variable body with a wide range of attributes. Perhaps the best we can do at this stage is to define "soil" as the naturally occurring fragmented, porous, and relatively loose assemblage of mineral particles and organic matter that covers the surfaces of our planet's terrestrial domains. Formed initially by the physical disintegration and chemical alteration of exposed rocks (a complex of processes called "weathering"), the soil is subsequently influenced by the activity and accumulated residues of diverse forms of life. It is within that natural body that living roots of plants can obtain anchorage and sustenance, alongside a varied interdependent biotic community of microscopic and macroscopic organisms. A mere fistful of soil may contain billions of microorganisms that perform vital interactive biochemical functions. Another intrinsic attribute of the soil is its sponge-like porosity and its enormous internal surface area. That same fistful of soil may actually consist of several hectares of active surface, upon which physicochemical processes take place continuously.

Considering the height of the atmosphere, the thickness of the earth's rock mantle, and the depth of the ocean, we note that soil is an amazingly thin body—typically not much more than one meter thick, and often less than that. Yet it is the incubator of terrestrial life within which biological productivity is generated and sustained. It acts as a composite living entity.

However unique in form and function, the soil is not an isolated body. It is, rather, a central link in the larger chain of interconnected domains and processes comprising the biosphere. The soil interacts both with the overlying atmosphere and the underlying strata, as well as with surface and underground bodies of water. Especially important is the interrelation between the soil and the climate. In addition to its function of regulating the cycle of water, the soil also regulates energy exchange and surface temperature. Through sequestration of carbon in the soil as organic matter, or, contrariwise, release of soil carbon as methane or carbon dioxide, the soil also has an effect on the composition and the radiative properties of the atmosphere. Hence, the soil has an effect on the mitigation or enhancement of the so-called "greenhouse effect"—the cause of global warming.

HUMANITY'S RELATIONSHIP TO THE SOIL

Realizing humanity's dependence on the soil, ancient peoples, who lived in greater intimacy with nature than many of us do today, revered the soil. It was not only their source of livelihood, but also the material they used to build their homes and that they learned to shape, heat, and fuse into bricks, household vessels, and writing tablets (ceramic—made of clayey soil—being the first synthetic material in the history of technology).

In the Hebrew Bible, the name assigned to the first human was "Adam", derived from "adama", meaning "soil". The name given to his mate was "Hava" (transliterated as "Eve"), meaning "living" or "mother of life". Together, therefore, "Adam and Eve" signify, quite literally, "soil and life". The same powerful metaphor is echoed in the Latin name for the human species—*Homo*, derived from "humus", the material of the soil. Hence, the adjective "human" also suggests "of the soil". Other ancient cultures evoked equally powerful associations. To the Greeks, the earth was a manifestation of Gaea, the maternal goddess who, impregnated by the sky-god Uranus, gave birth to all the gods of the Greek pantheon.

Our civilization depends on the soil more crucially than ever, because our numbers have grown while available soil resources have diminished and deteriorated. Around the world, we are losing more and more of our most productive soils to the encroachment of cities, industries, and roadways. We are also degrading the remaining soils by subjecting them to erosion, compaction, leaching of nutrients, loss of organic matter, acidification, salination, depletion of water supplies, and pollution.

Thousands of years are required for nature to create life-giving soil out of sterile bedrock. In only a few decades, unknowing or uncaring humans can destroy that wondrous work of nature. Mismanaged soils tend to erode, and their transported sediments clog streambeds, estuaries, lakes, and coastal waters. The runoff and leaching from such soils may cause waterlogging and impaired aeration of lowlands, eutrophication of streams and lakes, and accumulation of salts. Such processes of soil degradation already affect large areas of once-productive land.

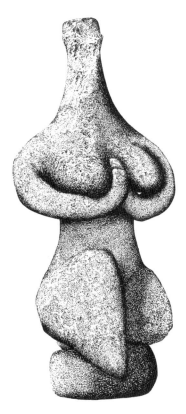

Fig. 1.1. Neolithic figurine from Tel Chagar Bazar, Mesopotamia, ca. 6000 BCE. The fertility of the earth was symbolized by the image of an ample goddess made of clay.

The optimal soil is deep and permeable, with balanced combinations of mineral and organic matter, fine and coarse particles, and pores having an array of sizes that contain and conduct water and air. But the soil is a labile and ever-vulnerable body. When its natural state is disrupted—that is to say, when it is bared of its protective vegetative cover and then subjected to the impact of raindrops and human manipulation (pulverization, compaction, and consequent erosion), as well as to chemical contaminants—the soil's fragile equilibrium is destroyed, and it undergoes degradation. Proper soil management implies using soils to their full productive potential in a sustainable manner so as to conserve—and even to enhance—their productivity in the long run while avoiding deterioration of the environment (air, water, adjacent soils, and biota). Where soils have been degraded in the past, the necessity is to find efficient and economical means to remediate them.

CULTURAL ATTITUDES TOWARD THE SOIL

Throughout history, the ways different societies used or abused their soil resources have influenced their ultimate fates. That fundamental fact is no less true today than in the past. Paradoxically, even as our dependence on the soil has increased, most of us have become physically and emotionally detached from it. Many people in so-called "developed" countries spend their lives in the artificial environment of a city, insulated from direct exposure to nature, and some city-bred children may now assume as a matter of course that food originates in supermarkets.

Detachment has bred ignorance, and out of ignorance has come the delusion that our civilization has risen above nature and has set itself free of nature's constraints. Agriculture and food security, erosion and salination, degradation of ecosystems, depletion and pollution of surface waters and aquifers, changes in atmospheric composition and the earth's energy balance, deforestation and decimation of biodiversity—all these processes, which involve the soil directly or indirectly—have become abstractions to many people. The very language we use conveys disdain for that common material underfoot, too often referred to as "dirt". Some fastidious parents prohibit their children from playing in the mud and hurry to wash their "soiled" hands when the children obey an innate instinct to do so. Thus, soil is devalued and treated as unclean even though it is the terrestrial realm's principal medium of purification, wherein wastes are decomposed and nature's productivity is regenerated.

We cannot manage effectively and sustainably so complex, vital, and labile a medium as the soil unless we understand its attributes, inner functions, and environmental interactions. That is why the tasks of developing and disseminating sound knowledge of the soil and its processes have assumed growing urgency and importance. The global environmental crisis has created a compelling need for educating the next generation of aspiring environmental scientists and activists—and through them, the general public—regarding the soil in all its manifestations, in nature and in relation to the life of humans.

The Latin word for "soil", "humus", by which we now refer to the organic fraction of the soil, gave rise not only to the name of our species, *Homo*, but also to the words "humble" and "humility", implying "lowly as the earth". It therefore seems ironic that we have assigned our species so arrogant a name as *Homo sapiens sapiens* ("wise wise earthling"). As we ponder our relation to the soil, we might consider adopting a more modest appellation, namely *Homo sapiens curans*, with the Latin adjective *curans* denoting "caring", or "caretaking" (as in the title "curator"). Of course, we must strive to deserve the new name, even though we have not always deserved the present one. The soil has always nurtured us, despite our abuse of it. We cannot continue to behave as its ungrateful children. It is time for us, as *Homo sapiens curans*, to nurture the soil in return.

2. SOIL IN THE HISTORY OF CIVILIZATION

*History is largely a record of human struggle
to wrest land from nature.*
H. H. Bennett and W. C. Lowdermilk

HUMAN ORIGINS IN RELATION TO THE ENVIRONMENT

Humanity's birthplace was evidently in the continent of Africa, and its original habitat was the subtropical savannas that constitute the transition zone of sparsely wooded grasslands lying between the zone of the humid and dense tropical forests and the zone of the arid steppes. We can infer the warm climate of their place of origin from the fact that humans are naturally so scantily clad, or furless; and we can infer the open landscape from the way humans are conditioned to walk, run, and gaze over long distances.

For at least 90% of its existence as a species, the human animal functioned merely as one member of a community of numerous species who shared the same environment. Early humans were adapted to subsist within the bounds defined by the natural ecosystem. By and large, our ancestors led a nomadic life, roaming in small bands, foraging wherever they could find food. They were gatherers, scavengers, and opportunistic hunters. Unlike their primate cousins who remained primarily vegetarian, humans diversified their diet to include the flesh of whichever animals they could find or catch, as well as a variety of plant products such as nuts, berries and other fruits, seeds, some succulent leaves, bulbs, tubers, and fleshy roots.

Hundreds of thousands of years ago, during the Paleolithic period, members of the *Homo* genus learned to fashion stone tools as well as to set and use fire, probably at first only to cook and soften food. Later, they used fire to clear woody vegetation, to flush out game, and to encourage the subsequent grazing of ungulates, which they could thereby hunt more readily. Eventually, that practice had a great effect on the environment, as brush fires apparently set by humans modified entire ecosystems on an increasing scale. The control of fire became even more important when humans moved out of the tropics into colder climes, where bonfires and hearths were needed to warm their shelters in winter, as well as to cook their food. In time, the clearing of woodlands and shrub lands by repeated firings also set the stage for the advent of agriculture. However, as the vegetative cover was affected by fire, so was the underlying soil. Following repeated denudations of the land, soil organic matter and nutrients were gradually depleted while soil erosion took place, resulting in the increased transport of silt by streams and its deposition in river valleys and estuaries.

THE AGRICULTURAL TRANSFORMATION

By the later stages of the Paleolithic period, nearly all the regions of human habitation had experienced some anthropogenic modification of the floral and faunal communities. The gradual increase in human population density required the development of more intensive methods of land use, aimed at inducing an area to yield a greater supply of human needs. The selective eradication of less desirable plant and animal species and the encouragement of desirable ones (producing edible and, preferably, storable products) led eventually to the domestication and propagation of crops and to purposeful soil management aimed at creating favorable conditions for production. That is to say, these activities culminated in the development of agriculture and the sedentary (rather than nomadic) way of life in villages.

Concurrently with the domestication of plants, humans also learned to select and domesticate animals. The earliest domesticates in the ancient Near East were sheep, goats, and bovine cattle, which could be herded and pastured in the semiarid rangelands. Cattle (oxen) were also used as beasts of burden in order to pull carts as well as to plow the soil. Donkeys, horses, and, eventually, camels were also harnessed in the same region to convey people and goods.

The Agricultural Transformation, also called the Neolithic Revolution, was probably the most momentous turn in the progress of humankind, and many believe it to be the real beginning of civilization. It first took place in the Near East approximately ten millennia ago, initially in the arc of rain-fed uplands and foothills fringing the so-called Fertile Crescent. Prominent among the native plant resources of that region were species of the *Graminea* and *Leguminosa* families (the former included the progenitors of wheat, barley, oats, and rye; the latter included lentils, peas, chickpeas, and vetch). The nutritious seeds of such plants could be collected and stored to provide food for several months. Most native plants scattered their seeds as soon as they mature and, therefore, were difficult to harvest efficiently. A few anomalous plants, however, due to chance mutations, retained their seeds. The preferential selection of such seeds, and their propagation in favorable plots of land, constituted the beginnings of agriculture, providing the early farmers with crops that could be harvested more uniformly and dependably than could the wild plants.

Eventually, farming was extended from the relatively humid centers of its origin to the river valleys of the Jordan, the Tigris-Euphrates, the Nile, and the Indus. The climate of these valleys was generally arid, but they received the excess runoff via rivers flowing in from watersheds in the rainfed highlands. Here a new type of agriculture, based primarily on irrigation, came into being. The early farmers of those river valleys, around 5000 B.C.E., relied at first on natural irrigation by the unregulated seasonal floods to water the banks or floodplains of the rivers. As soon as a flood withdrew, they could cast their seeds in the mud. At times, however, the flood did not last long enough to wet the soil thoroughly, so the riparian farmers learned to build dikes around their plots to create basins in which a desired depth of water could be impounded until it soaked into the ground and wet the soil fully. The basins also retained the nutrient-rich silt and prevented it from running off with the receding flood-waters. Channels were dug to convey river water to the farther reaches of the plain that might not receive the water otherwise. By these means, irrigation farmers strove to gain greater control over the water supply to their crops.

Rain-fed and/or irrigated farming practices spread to, or were developed independently in, other human-inhabited regions around the world, including India, China, and Mesoamerica.

The ability to raise crops and livestock, while resulting in a larger and more secure supply of food than was previously possible, definitely required attachment to controllable sections of land and hence brought about the growth of permanent settlements and of larger coordinated communities. The economic and physical security so gained accelerated the process of population growth and necessitated further expansion and intensification of production. A self-reinforcing pattern thus developed, so the transition from the nomadic hunter-gatherer mode to the settled farming mode of life became, in effect, irreversible.

The Agricultural Transformation radically changed almost every aspect of human life. Food production and storage stimulated specialization of activities and greatly enhanced the division of labor that had already started in hunting-gathering societies. The larger permanent communities based on agriculture required new forms of organization, both social and economic in nature. Domestication affected family structure and the roles and statuses of men, women, and children. With permanent facilities—such as dwellings, storage bins, heavy tools, and agricultural fields—there came the concept of property. The inevitably uneven allocation of such property resulted in self-perpetuating class differences. Religious myths and rituals, as well as moral and behavioral standards, developed in accordance with the re-formed economic and social constellation and the changed relationship between human society and the environment.

SOIL AND WATER HUSBANDRY AND CERAMICS

An important factor in the evolution of agriculture in the Near East, as elsewhere, was the development of the tools of soil husbandry. Seeds scattered on the ground were often eaten by ants, birds, and rodents, or subject to desiccation, so their germination rate was likely to be very low and uneven. Given a limited seed stock, farmers would naturally do whatever they could to promote germination and seedling establishment. The best way to accomplish this was to insert the seeds to some shallow depth, under a protective layer of loosened soil, and to eradicate the weeds that might compete with the crop seedlings for water, nutrients, and light.

The simplest tool developed for the purpose was a paddle stick, by which a farmer could make holes for seeds. The use of this simple device was extremely slow and laborious, however, so at some point the digging stick was modified to form the more convenient spade, which could not only open the ground for seed insertion but could also loosen and pulverize the soil and eradicate weeds more efficiently. In time, the space developed a triangular blade, at first made of wood, later made of stone, and eventually made of metal. The spade, initially designed to be used by one person, was later modified so that it could be pulled by a rope, so as to open a continuous slit, or furrow, into which the seeds could be sown. A second furrow could then be made alongside the first to facilitate weed coverage. In some cases, the rows were widely enough separated to permit a person to walk between the rows and weed the cultivated plot.

The human-pulled traction spade (called an *ard*) gradually metamorphosed into an animal-drawn plow. The first picture of such a plow, dating to 3000 B.C.E., was found in Mesopotamia, and numerous later pictures have been found both there and in Egypt, as well as in China. It was not long before these early plows were fitted with a seed funnel, so that the acts of plowing and sowing could be carried out simultaneously. The same ancient implement is still in use today in parts of the Near East and Africa.

Although the development of the plow represented a great advance in terms of convenience and efficiency of operation, it had an important side effect. As with many other innovations, the benefits were immediate, but the full range of consequences took several generations to play out, long after the new practice became entrenched. The major environmental consequence was that plowing made the soil surface—now loosened, pulverized, and bared of vegetation—much more vulnerable to erosion.

As the population grew, especially in the irrigated river valleys, the necessity arose to intensify production still further. Instead of just one crop per year, the farmers of Mesopotamia and Egypt, for example, could grow several crops,

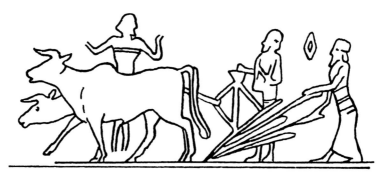

Fig. 2.1.   Depiction of a seeder-plow on an ancient Mesopotamian seal.

given the year-round warmth and abundant sunshine of the local climate. To do so, they needed to draw water at will from the rivers or from shallow wells dug to the water table. At first, they drew water manually in animal skins or ceramic containers, which they then carried using shoulder yokes. In time, a new technology was invented: mechanical water-lifting devices. The simplest of these is the device known today in Arabic as the *shadoof*—a wooden pole used as a lever, with the long arm serving to raise bucketfuls of water and the short arm counterweighted with a mass of mud or a stone. A more sophisticated device was invented much later and attributed to Archimedes, the Greek scientist who discovered the principle named after him (that bodies immersed in water are buoyed by the weight of the water they displace). His device consisted of an inclined tube containing a tight-fitting spiral fin. Both the shadoof and Archimedes' screw devices were human powered. A more elaborate water-lifting device was the animal-powered waterwheel. All these devices were still in use until quite recently; as of late, they have been replaced by modern motorized pumps.

As farming induced sedentary living in villages, there also developed an important new industry that depended directly on the soil—ceramic pottery, which began in the Near East at approximately 6000 B.C.E. The shaping and baking of clay to form hardened vessels for grain, for liquid storage and conveyance, and for cooking, represented the first transmutation of material by

Fig. 2.2. Water-lifting devices in ancient Mesopotamia and Egypt (adapted from ancient murals).

humans. Such an innovation could not have been possible, owing to the fragility of the ceramic objects, during the nomadic hunting-gathering phase of human subsistence. Bricks made of molded clay served for the construction of homes and grain storage, while tablets of moist clay could be imprinted with various signs and then hardened in a kiln to form permanent records in an early form of writing called *cuneiform*. Later still, people learned to extract, alloy, and mold metals such as copper and iron and to fashion them into implements.

LAND DEGRADATION

Manipulation and modification of the environment was a characteristic of many societies from their very inception. Long before the advent of earth-moving machines and toxic chemicals, even before the advent of agriculture, humans began to affect the land and its biota in ways that tended to destabilize natural ecosystems.

The evolution of agriculture left an ever-stronger imprint on the land in many regions. The vegetation, animal populations, slopes, valleys, and soil cover of land units were radically altered. The practices of tillage and fallowing, terracing, irrigation and drainage, as well as grazing of flocks, further accelerated soil erosion in sloping lands and sedimentation in lowlands. Soil lost from denuded cultivated slopes could not be regenerated unless the land was allowed to revert to its vegetative cover for many decades or centuries. Moreover, soils that were irrigated in poorly drained river valleys tended to become waterlogged and saline, with the result that the practice of farming there could not be sustained in the long run.

In many of the ancient countries, where human exploitation of the land began early in history, we find disturbing examples of once-thriving regions reduced to desolation by human-induced degradation. Some of the early civilizations succeeded all too well at first, only to set the stage for their own eventual demise. The poor condition of the environment in the Fertile Crescent today is not simply due to a changing climate or to devastation caused by repeated wars, although both of these may well have had important effects. It is due in large part to the prolonged exploitation of this fragile environment by generations of forest cutters and burners, grazers, cultivators, and irrigators, all diligent and well-intentioned, but destructive nonetheless.

The same anthropogenic processes that began in the early history of civilization have continued ever since, on a more extensive scale. Especially vulnerable are cultivated and overgrazed soils in semiarid regions, where droughts occur periodically, accompanied by strong winds that deflate the bare and pulverized soil surface, and where the intermittent onset of rains causes further damage. Consequently, once-productive lands may become so degraded as to acquire desert-like characteristics. Hence the process of land degradation in semiarid regions has been termed *desertification*.

Examples of large-scale land degradation can be cited in many parts of the world, even up to the present. A case in point is the famous "Dust Bowl" that took place in the Southern Great Plains of the United States in the 1930s and that caused the uprooting and migration of entire farming communities.

A similar process has been occurring on a vaster scale in the African Sahel, which is the continent-wide semiarid savanna belt that lies between the Sahara Desert in the north and the tropical forests in the south. A particularly disastrous example of water-logging and salination due to irrigation is seen in the Aral Sea Basin of Central Asia (Uzbekistan, Turkmenistan, and Kazakhstan). The same processes are occurring in highly developed parts of the world, such as the Murray-Darling Basin of Australia, as well as the San Joaquin Valley of California. Even more consequential is the large-scale destruction of rainforests and the loss of biodiversity resulting from the expansion of agriculture in the tropical zones of South America, Central Africa, and Southeast Asia, where the warm temperatures and high humidity contribute to the rapid decomposition of the soil's original organic matter content and to the deterioration of the soil's structure as well as its nutrient content.

A growing awareness of the dire consequences of land degradation has led to the development of new methods of soil management, designed to conserve and even to enhance soil quality in the long run. Such methods are based on minimizing the mechanical disruption and the chemical contamination of the soil, and on maximizing its biodiversity and organic matter content.

He gathered some of the earth up in his hand and he sat and held it thus, and it seemed full of life between his fingers. And he was content, holding it thus.
Pearl Buck, *The Good Earth* (1935)

BOX 2.1 Soil, Sustainability, and Society

The concept of sustainable use implies that a resource is utilized in such a manner as to preserve its quality and productive potential; in other words, that its use for our needs at present will not diminish the quality of the resource and the ability of future generations to continue using it for their own needs. In practice, ensuring its sustainability requires that the soil be protected against any and all processes of degradation. Those processes may include one or more of the following: erosion by water or wind; acidification or alkalization; salination; depletion of nutrients by leaching and/or crop extraction without replenishment; loss of organic matter by decomposition and gaseous emission; deterioration of soil structure (deflocculation of clay, compaction, breakdown of aggregates, excessive pulverization by tillage); contamination by toxic or pathogenic factors; restriction of aeration; and hydrophobization (water repellency).

Especially fragile are the highly weathered soils of the humid tropics, in which soil nutrients exist mainly in organic form (i.e., in plant and animal residues that decompose rapidly in warm and wet conditions) rather than in the more stable mineral form that is typical of less-weathered soils. Also fragile are the soils of arid regions, where tillage or overgrazing exposes the soil's surface to periods of drought and deflation by wind, alternating with short spurts of intensely erosive rainstorms.

Once degraded, soils are generally difficult, though not necessarily impossible, to rehabilitate. Productivity may be restored and even enhanced by better modes of watershed management, judicious additions of nutrients and organic matter, conservation of soil and water, and—where necessary and possible—efficient irrigation. Finally, productive soils must be protected from unbridled urban expansion (which real-estate agents deceptively call "development").

3. SOIL FORMATION

There is nothing Mother Nature loves
so well as to change existing forms
and to make new ones like them.
Marcus Aurelius Antonius (121–180)

WEATHERING

The phenomenon of weathering, consisting of the physical fragmentation of exposed rocks and the chemical decomposition and recomposition of their mineral constituents, initiates the complex process of soil formation. But how can the seemingly feeble forces of the weather bring about the breakdown of strong, dense, and massive bedrocks such as granite or basalt? The answer is that those forces act slowly but relentlessly.

Exposed rock surfaces expand and contract repeatedly as they are subjected to fluctuations of temperature from day to night and from summer to winter. Additional stresses occur due to the wetting and drying of porous rocks, and the freezing and thawing of water in cracks. Growing roots penetrate these cracks and exert powerful swelling pressures that further split the rocks. Abrasive particles carried by creeping ice, flowing water, and blowing wind scour and grind rock surfaces. In addition, soluble components of minerals present in the rocks are dissolved and transported by recurrent rain, percolate into the fragmented rock, and re-precipitate at various depths.

Superimposed upon these physical processes and augmenting them are numerous chemical reactions that tend to modify still further the minerals present in the original bedrock. Among these chemical processes are hydration, oxidation and reduction, ionic dissociation of dissolved electrolytes, acidification, and alkalization.

The selective removal of various components occurs through dissolution, leaching, precipitation, and volatilization. Although some *primary minerals*—present in the bedrock—are relatively stable and retain their original character even in fragmented form (e.g., quartz), other minerals are quite reactive (e.g., feldspars and micas). The latter minerals gradually decompose and are reconstituted into a series of secondary minerals, formed within the soil. Prominent among these *secondary minerals* is a very distinctive group called *clays*. We shall have much more to tell about clays in subsequent sections of this book.

However, the collection of physically fragmented and chemically altered material is not yet a true soil, just as a random collection of bricks of sundry sizes, shapes, and compositions, thrown haphazardly into a heap, does not yet make a building. The same bricks, only differently arranged and mutually bonded, can form a home or a factory. Similarly, an assemblage of weathered rock fragments must undergo further development before it can be considered a true soil. The full evolution of a soil may require many centuries or even millennia.

SOIL FORMING FACTORS

Five principal factors influence the genesis of, and determine the eventual character of, a distinctive soil: These factors act in concert: parent material, climate, biotic community, topography, and time.

Parent Material

The first factor to consider is the nature of the *parent material*. Soils that form directly over *in situ* bedrock are called *residual* soils. Geologists recognize three main types of rocks: *igneous*, congealed from molten magma; *sedimentary*, consolidated and often cemented sediments; and *metamorphic*., either igneous or sedimentary rocks that have subsequently been modified by heat and pressure. Alternatively, soils may develop from non-residual (i.e., transported) unconsolidated material, such as *alluvial* (water-laid), *aeolian* (wind-laid), or *glacial* deposits.

Each of the numerous types of parent material has its innate physicochemical characteristics that influence soil formation from the outset. For example, soils derived from granite typically contain quartz sand and tend to be acidic, whereas soils derived from basalt are often clayey and chemically neutral or alkaline.

The effect of the parent material is most apparent in "young" soils, in which the minerals have not yet undergone complete transformation. Over a long period of time, the originally present minerals may be transmuted into secondary minerals, reflecting the action of the various dynamic soil-forming factors, which eventually bring about the formation of "mature" soils. Some soils are kept perpetually young by the constant or repeated deposition of new sediment (laid down by water or by wind), or by the continuous removal of topsoil and the exposure of freshly weathered bedrock.

Climate

A second and—in the long run, often predominant—factor of soil formation is the climate. Temperature and moisture affect the rates of dissolution,

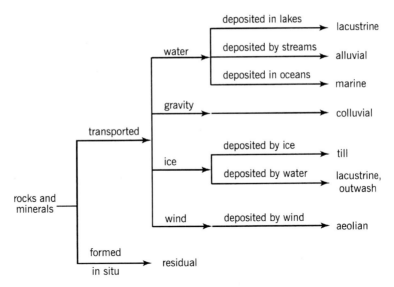

Fig. 3.1. Classification of parent materials. (Dubbin, 2001).

chemical decomposition, leaching, or deposition of soil components. In warm and humid regions, readily soluble constituents (such as the basic cations Ca, Na, K, and Mg) may be almost totally washed out of soils, leaving behind only resistant minerals such as iron oxides. Such soils tend to be less fertile, due to the excessive loss of soluble nutrients.

On the other hand, in warm but relatively dry (semiarid) regions, calcium carbonate or sulfate may precipitate (rather than be leached) and accumulate to form a cemented ("indurated") layer at some depth in the soil. In still drier regions, even such highly soluble salts as sodium chloride may be retained in the soil. Organic matter—the partially decomposed residue of plants and animals—tends to accumulate in relatively cool climes, but to decompose rapidly in warmer ones.

The effect of climate is seen most strongly where a region spanning a gradation of climatic conditions (from wet to dry and from warm to cold) is underlain by uniform parent materials. Such was the condition in Russia, where the dominant role of climate on soil formation was first noticed.

Mature soils that are more or less in equilibrium with the prevailing regional climate are said to be *zonal*. Where the soil in a given climatic zone still retains some influence of the parent material, it is called *intrazonal*. However, where the processes of soil formation have been strongly restricted by local conditions and the soil maintains the major characteristics of its parent material, it is termed *azonal*.

Biotic Community

A third factor of soil formation is due to the nature and activity of the biotic community inhabiting the soil, which itself is reciprocally conditioned by the climate and nature of the substrate.

Vegetation has a strong influence on soil formation. Soils formed under deciduous forests differ from soil under grasslands, even where the climate is not radically different. Generally, however, the character of the predominant vegetation is itself dictated by the climate. Soils that develop in a relatively humid region under a cover of forest trees tend to accumulate organic matter at the surface but often exhibit a leached layer underneath. In contrast, soils formed under a grass cover tend to accumulate organic matter throughout a deeper layer.

Other organisms that affect the soil are various soil-inhabiting animals such as earthworms, termites, and rodents that dig in the soil and mix it to some depth. They thus help to convey organic matter to greater depth and to create channels that facilitate the movement of water and air, as well as the migration of particles and solutes.

Even more profoundly influential are the various microorganisms such as fungi, protozoa, and bacteria (including the filamentous and rod-shaped types known as *actinomycetes*), which process organic matter either aerobically or anaerobically. Depending on the soil's moisture content and rate of aeration (i.e., the diffusive supply of oxygen and the release of carbon dioxide via the soil's air-filled pore spaces), soil microorganisms of one group or another either oxidize or chemically reduce organic matter and minerals.

Topography

A fourth factor of soil formation is the configuration of the landscape; i.e., the topography of the area in which the soil develops.

Topography affects soil formation in various ways. Where the land is flat, the processes of energy exchange and of water inflow and release tend to be vertical, so the soil develops to a characteristic depth. In contrast, where the land slopes steeply, a considerable portion of the rainfall flows downslope over the surface (a phenomenon called *runoff*), often scouring the surface and causing erosion. Consequently, the soils on sloping ground tend to be shallower and drier that those situated on plateaus or in valleys.

The water shed from the sloping ground brings more moisture and deposits additional sediment in the valleys, or bottomlands. Valley soils may even accumulate shallow groundwater due to impeded drainage, and consequently be poorly aerated.

The soils that form in sequential sections of the landscape tend to differ in microclimatic conditions, although they are located in the same macroclimate zone and on similar parent material. The succession of such soils—from plateau or hilltop to slope to hill bottom to valley—is called a *toposequence*, or *catena* (from the Latin word suggesting "a chain").

Time

The fifth factor of soil formation is the duration, or time, over which the soil forming processes have been active. Time is a factor of soil formation because the various processes that eventually determine the characteristics of the soil act gradually and their influence on soil formation may culminate in the establishment of a mature soil only after some centuries or millennia.

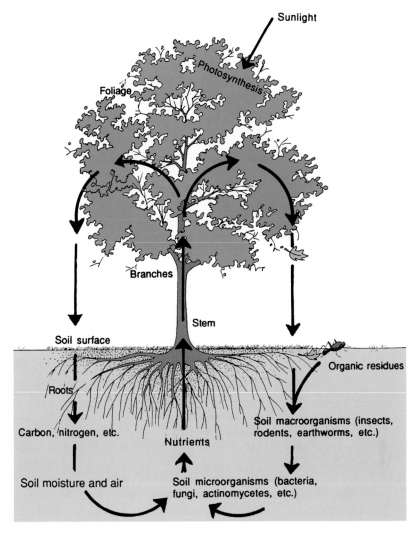

Fig. 3.2. The role of the biotic community in soil formation (schematic).

The rate of soil formation is influenced by prevailing conditions of moisture and temperature. Warm, humid conditions hasten, whereas cool and dry conditions inhibit, weathering and soil formation. Where those processes have persisted for an extended period, the resulting soils may have attained a practically stable character.

In the course of time, the original nature of the parent material becomes less important as the soil assumes a character most strongly influenced by the climate (which in turn determines the biochemical and biophysical processes and their ultimate expression). The climate also determines the intensity of leaching. While soils in humid regions typically lose all readily soluble salts, soils in arid regions tend to accumulate salts that are not leached, or that rise wherever the soil solution is drawn by capillary action toward the soil surface. There the water evaporates and the salts are left behind and tend to accumulate.

Over a very long period, soil formation may not end in a truly stable "equilibrium" condition, but may even lead to a self-degraded "senile" condition (such as a soil with an impermeable substratum, or a completely waterlogged condition). On the other hand, recent deposits that have been subject to soil formation for relatively short periods (hence, the processes are incomplete) are considered to be young, or immature, soils. Young soils, formed on either residual or on alluvial parent material, are often more fertile—having retained soluble nutrients—than old soils that have been thoroughly leached of nutrients.

*

At this point, a caveat is in order. The five factors listed are not mutually independent. Rather, they are intertwined. The biotic community, for instance, depends on the prevailing climate, just as a soil's own microclimate—its moisture and temperature regimes—depends on its position in the landscape—the topography—as well as on the nature of the underlying parent material—whether it is permeable or relatively impervious. The parent material itself is often related to the topography in that hillslopes may expose bedrock, whereas valleys are typically blanketed with sediment.

Human Intervention

A sixth factor that has gained importance over time is *human intervention*. It includes the clearing of native vegetation and the resulting disruption of entire natural ecosystems; reshaping the land surface and modifying its drainage pattern; baring and mechanically cultivating topsoil and the consequent compaction, disruption of soil structure, erosion, and loss of organic matter; seasonal cropping patterns and removing biomass; producing changes by the application of pesticides, fertilizers, irrigation, and drainage; and possibly degrading the soil by salination and chemical—possibly toxic—pollution. Not all anthropogenic influences, however, are necessarily destructive. In some places, judicious management of soil fertility, organic matter, and soil moisture may actually enhance soil productivity and biodiversity. Consequently, managed soils may differ markedly from the original or natural soils. Sadly, however, cases where human intervention has resulted in enhanced productivity are fewer and less extensive than cases where it has caused soil degradation.

SOIL FORMATION: DEVELOPMENT OF THE SOIL PROFILE

Most of the soils we observe today have been formed during the current climate era known as the Holocene. Earlier cycles of soil formation have been largely obscured by erosion—especially in northerly latitudes—due to glaciation, rainfall, flooding, or wind. In some areas, however, older cycles of soil formation can be observed, covered by sediment layers of subsequent deposition.

Chemical weathering, decomposition, and recomposition depend on the presence of water and on the prevailing temperature. In arid environments,

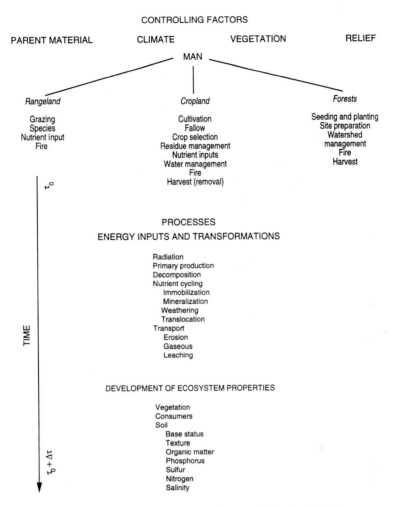

Fig. 3.3. Soil-forming factors and processes. (Coleman and Crossley, 1996).

physical weathering occurs with little chemical change, so the products are largely unaltered fragments of rocks and minerals. In humid environments, the temperature determines the rate of decomposition, and it can also influence the nature of the products that are formed.

Soil formation consists of a gradual and progressive series of processes. As soon as the exposed rock begins to fragment and to form a porous outer layer containing soluble nutrients, that substrate is permeated by water and air and is colonized by plants, whose organic residues in turn are decomposed by microorganisms. Fauna such as earthworms then enter the fray, burrowing and churning the loose material, opening channels, and mixing the organic and mineral components. In the course of chemical decomposition and recomposition, some of the original minerals are transformed into clay, which swells and shrinks alternately and adsorbs ions. Clay particles tend to clump together to form flocs, which, when cemented by organic colloids, form larger aggregates.

Water percolating through the interstices leaches some of the soluble components released by mineral weathering and organic-matter decomposition, and carries them deeper into the profile. Additional components are contributed by deposition of gases and particulates from the atmosphere. Acids brought by rain or released by organic decomposition further solubilize soil minerals.

In humid conditions, where rainfall exceeds evaporation, the solutes tend to be leached downwards to the groundwater aquifer, and then they drain toward rivers and the sea. Where the evaporation rate exceeds rainfall, at least during part of the time, the solutes may reverse course and rise with capillary water toward the surface. Such is the case in arid regions, where salts may accumulate within, and at the surface of, the soil.

As percolating water carries solutes, it may also carry particles in suspension. The upper part of the soil—i.e. the A horizon, or topsoil—from which the fine particles of clay are "washed out" is called the *eluvial* zone. In some soils, the A horizon, just below the organic-darkened surface, is so impoverished of clay as to seem bleached, a condition called *podzol* ("ash-like") in Russian.

The lower part of the soil into which the clay is carried and in which it tends to lodge—the B horizon, or subsoil—is called the *illuvial* zone. (*Eluviation* and *illuviation* imply "washing out" and "washing in," respectively.) Soil aggregates in the B horizon are often covered with a coating of clay, called *clay skins*. In some cases, the clay-enriched subsoil may become so clogged as to be relatively impermeable, a condition known as a *claypan*.

In extreme cases of limited aeration, the oxidation of organic matter is so slow as to allow the accumulation of organic matter in the form of peat. Iron is reduced from Fe^{3+} to Fe^{2+}, and the soil loses the typically reddish or brownish color of ferric oxide and becomes grayish. The soil is then characterized as *gleyed*. Where the restriction of aeration is intermittent rather than continuous, the chemical reduction is only partial, and the soil may exhibit mottling, with alternating spots or splotches of grey and brown. The interiors of aggregates are particularly likely to undergo reduction, while the exteriors remain oxidized.

∗

In general, the process of soil formation consists of three sequential stages, which tend to blend into one another: (1) weathering of the parent material; (2) formation of clay and accumulation of organic matter; and (3) translocation of matter and differentiation of horizons. The entire process culminates eventually in the establishment of a characteristic soil profile, which manifests the variation of soil properties in depth. As different combinations of soil-forming factors in different environments give rise to distinctive soil types, the profile of each type is specific.

A soil profile is a vertical cross section of the soil that can be viewed by excavating a rectangular pit or sampling the soil layer by layer. The soil profile typically consists of a succession of more-or-less distinct strata. In the case of wind-deposited (aeolian) or water-deposited sediments, the strata may result from the pattern of deposition of the parent material.

If, however, the strata develop in place by internal soil-forming (*pedogenic*) processes, they are called *horizons*: The upper horizon is called the *A horizon* (or, in popular parlance, *topsoil*), the underlying one is called the *B horizon* (the *subsoil*), and the relatively undifferentiated zone beneath is called the *C horizon* (the *parent material*).

To a casual observer, the most obvious part of any soil is its surface zone. Through it, matter and energy enter or depart from the soil. The surface may be smooth or pitted, massive or granular, continuous or cracked, hard or friable, level or sloping, bare or mulched, and fallow or vegetated. Surface conditions strongly affect processes such as infiltration of water and diffusion of gases, as well as germination and growth of plants.

The A horizon is the zone of major biological activity. Here plants and animals and their residues interact with an enormously diverse multitude of microorganisms, billions of which may exist in a mere handful of topsoil. The topsoil is therefore generally enriched with organic matter, which consists of residues of plant and animal products at various stages of decomposition. The relatively stable fraction of the organic matter, called *humus*, typically imparts a dark hue to the A horizon.

In its natural state, the A horizon is generally the most fertile part of the soil, rich in nutrients as well as in organic matter. Once cultivated, however, the topsoil becomes the most vulnerable to degradation: It is often trampled, compacted, pulverized, bared of vegetative cover, and subject to the direct impact of rainstorms and winds. Consequently, the A horizon's originally high organic matter content is often depleted and its natural fertility and good structure tend to deteriorate.

The structure of the A horizon is typically crumb-like or granular, but in some special cases may exhibit a tendency to form a tight crust. The thickness of the A horizon varies from a few centimeters to a few decimeters. Some soluble components are typically leached from the A horizon by water percolating downward; hence, it is described as the *zone of eluviation* ("washing out"). When the surface of the A horizon is bared of a protective cover (e.g., vegetation or its residues), and especially when it is pulverized by cultivation, it becomes vulnerable to erosion, such as scouring by raindrop impact and runoff, and—when dry—deflation by wind.

Underlying the A horizon is the B horizon, where some of the labile materials, such as soluble minerals and migrating clay particles that are leached from the A horizon by percolating water tend to accumulate. Hence, the B horizon is described as the *zone of illuviation* ("washing in"). Those accumulations and the pressure of the overlying soil may combine to reduce the porosity of the deeper layer. In some cases, an overly dense B horizon may inhibit aeration, slow the internal drainage of water, and resist the penetration and proliferation of roots.

The clay-enriched B horizon is generally thicker than the A horizon, contains more clay but much less organic matter and biotic activity, and its structure is often massive—sometimes exhibiting columnar features—rather than loosely granular.

Underlying the B horizon is the C horizon, which is the soil's parent material. In a *residual soil* (i.e., a soil formed from *in situ* bedrock rather from pre-transported sediment), the C horizon consists of the partially weathered and fragmented rock material—a transition zone between the soil above and the original bedrock below. In soils that are not residual, the C horizon may consist of various pre-transported deposits of alluvial, aeolian, or glacial origin.

A hypothetical soil profile is illustrated in Fig. 3.3. This is not a typical soil, for among the myriad of greatly differing soils recognized by pedologists,

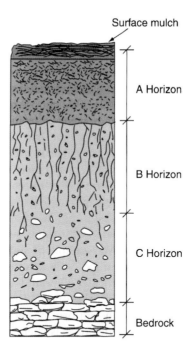

Fig. 3.4. Schematic representation of a hypothetical soil profile, showing: A horizon with aggregated crumblike structure; B horizon with columnar structure; and C horizon with partly weathered rock fragments.

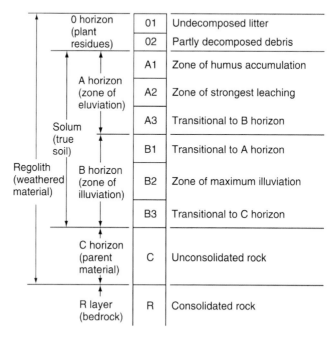

Fig. 3.5. Descriptive terminology for soil profile horizons.

no single soil can be considered "typical." The illustration is only meant to suggest the sort of contrasts in appearance and structure between successive depth horizons that may be encountered in a soil profile. (These contrasts will become evident in the depictions of different soil types presented in our next chapter.)

The A, B, C sequence of horizons is clearly recognizable in some cases, most generally in a typical *zonal* soil; i.e., a soil associated with a distinct climatic zone. In other cases, no clearly developed B horizon may be discernible, and the soil is then characterized by an A, C profile. In still other cases—for example in very recent alluvial deposits—hardly any profile differentiation is apparent. So-called *mature soils* are those that have been subjected to the factors of soil formation for a sufficient length of time for full profile development to have taken place.

PEDOGENESIS: SUMMARY

The systematic study of soil formation and classification is called *pedology*, and the process by which soils develop is called *pedogenesis*.

The typical development of a soil and its profile can be summarized as follows: It begins with the physical disintegration of an exposed rock formation, which provides the soil's parent material. Gradually, the loosened material is colonized by living organisms. The consequent accumulation of organic residues at and below the surface brings about the development of a discernible A horizon. This horizon may acquire an aggregated structure, stabilized to some degree by the cementing effect of the organic polymers. Continued chemical weathering (decomposition and recomposition) may bring about the formation of clay. Some of the clay thus formed tends to migrate downward, along with other transportable materials (such as soluble salts). The clay tends to accumulate in an intermediate zone (called the B horizon) between the surface zone of major biological activity and the deeper parent material of the so-called C horizon.

Important aspects of soil formation and profile development are the twin processes of eluviation and illuviation ("washing out" and "washing in," respectively), whereby clay and other substances emigrate from the overlying eluvial A horizon and accumulate in the underlying illuvial B horizon. The two horizons come to differ substantially in composition and structure. Throughout these processes, the profile as a whole deepens as the upper part of the C horizon is gradually transformed. Eventually, a quasi-stable condition is approached in which the counteracting processes of soil formation and of soil erosion are more or less in balance. In the natural state, the A horizon may have a thickness of 0.2 –0.4 m. When stripped of vegetative cover and pulverized or compacted excessively by tillage or trampling, this horizon becomes vulnerable to accelerated erosion by water and wind and may thereby lose half or more of its original thickness within a few decades.

In arid regions, salts dissolved from the upper part of the soil, such as calcium sulfate and calcium carbonate, may precipitate at some depth to form a cemented "pan" sometimes called *caliche*. Erosion of the A horizon may bring the B horizon to the surface. In extreme cases, both the A and B horizons may be scoured off by natural or human-induced erosion, so that the C horizon becomes exposed and a new cycle of soil formation may then begin. In other cases, a mature soil may be covered by a layer of sediments (alluvial or aeolian), so that a new soil forms over a "buried" old soil. Where episodes of deposition occur repeatedly over a long period of time, a sequence of soils may be formed in succession, thus recording the pedological history—called the *paleopedology*— of a region, including evidence of the climate and vegetation that had prevailed at the time of each profile's formation.

Numerous variations of the processes described are possible, depending on local conditions. The characteristic depth of the soil, for instance, varies from location to location. Valley soils are typically deeper than hillslope soils, and the depth of the latter depends on slope steepness. In places, the depth of the profile can hardly be ascertained, as the soil blends into its parent material without any distinct boundary. However, the zone of major biological activity seldom extends below 2–3 m and in many cases is shallower than 1 m.

4. SOIL CLASSIFICATION

All that we did, all that we said or sang
must come from contact with the soil...
William Butler Yeats (1865–1939)

ECOZONES OF THE WORLD

An overview of the globe's terrestrial domain reveals several distinct ecological zones, each of which manifests a specific combination of—and interaction among—factors of geology, topography, and climate (including radiation, temperature, and precipitation), which together constitute a formative environment for the evolution of a characteristic soil type and biologic community. At least nine ecological zones are recognized (Schultz 2005). They occur in bands, which may be fragmented owing to the irregular configuration of the continents and oceans, from the poles to the equator. Consequently, their boundaries are seldom sharp. The quantitative data regarding their sizes and distributions must therefore be regarded as rough estimates.

The various ecological zones differ greatly in their moisture and temperature regimes, as well as in their growing season lengths. In addition, they differ in the nature of their geologic substrata. These differences are reflected in the nature and intensity of their biological activity, expressible in terms of their net primary productivity (NPP)[1], and consequently in the distribution of their total biomass per unit area.

---

[1] Net primary productivity (NPP) of an ecosystem is defined as the amount of solar energy that is fixed biologically by photosynthetic plants (the so-called "primary producers"), minus the amount of energy these plants consume in respiration. As such, NPP provides the basis for the maintenance and growth of all species of consumers and decomposers. It is thus the total food resource on earth. In principle, as humans take more of this resource, less remains for other species.

Table 4.1 **Ecozones of the world and their areal extents (After Schultz, 2005)**

Ecozone	Area (Mkm2)	Landmass %	Growing season (mos)	Precipitation (mm/yr)
Polar-Subpolar (ice deserts, tundra)	22.0	14.8	0–4	100–250
Boreal (lichens, coniferous forests)	19.5	13.1	4–5	250–500
Temperate Midlatitudes (woodlands, forests)	14.5	9.7	6–12	500–1000
Dry Midlatitudes (grass steppes, semideserts)	16.5	11.1	0–4	< 400
Subtropics, Winter Rains (shrubs, conifers, broadleafs)	2.5	1.7	6–9	500–1000
Subtropics, Year-Round Rains (shrubs, sclerophyllous forest)	6.0	4.0	12	1000–1500
Dry Tropics and Subtropics (grass and shrub steppes)	31.0	20.8	0–4	250–500
Tropics, Summer Rains (dry and moist savannas)	24.5	16.4	6–9	500–1500
Tropics, Year-Round Rains (tropical rainforests)	12.5	8.4	12	2000–4000
Total area	149.0	100.0		

SOIL CLASSIFICATION SCHEMES

Given its wide range of differing ecological zones, the Earth naturally exhibits a highly diverse array of soil types that vary greatly in form and function. To make sense of that diversity, soil scientists have long sought to devise coherent systematic criteria by which to classify soils according to their properties and potential uses. The branch of soil science dealing with the characterization, classification, and geographic distribution of soils as they appear in nature is called *pedology*.

The earliest systematic approach to this science was developed by Russian pedologists in the late nineteenth and early twentieth centuries. Their work has subsequently been revised and augmented by pedologists from different regions and following different approaches. Currently, the two comprehensive classification schemes that appear to be in widest use are: (1) The system of soil classification developed jointly by the UN Food and Agriculture Organization (FAO) and the UN Educational, Scientific, and Cultural Organization (UNESCO); and 2) The system of *Soil Taxonomy* developed by the United States Department of Agriculture (USDA).

These schemes tend to group the soils of the world on the basis of the climatic factors that, in the main, influence their formation (called *pedogenesis*). However, the scheme recognize that specific local factors may impart special characteristics to soils that deviate to some degree from the principal climatological

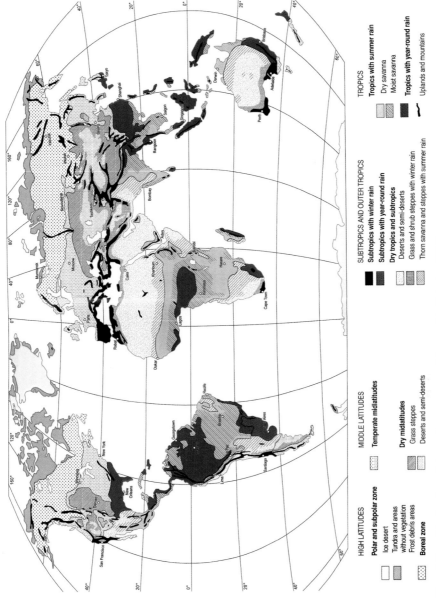

Fig. 4.1. Distribution of the major ecozones of the world. (Schultz, 2005).

HIGH LATITUDES

Polar and subpolar zone

Ice desert

Tundra and areas
without vegetation

Frost debris areas

Boreal zone

MIDDLE LATITUDES

Temperate midlatitudes

Dry midlatitudes

Grass steppes

Deserts and semi-deserts

SUBTROPICS AND OUTER TROPICS

Subtropics with winter rain

Subtropics with year-round rain

Dry tropics and subtropics

Deserts and semi-deserts

Grass and shrub steppes with winter rain

Thorn savanna and steppes with summer rain

TROPICS

Tropics with summer rain

Dry savanna

Moist savanna

Tropics with year-round rain

Uplands and mountains

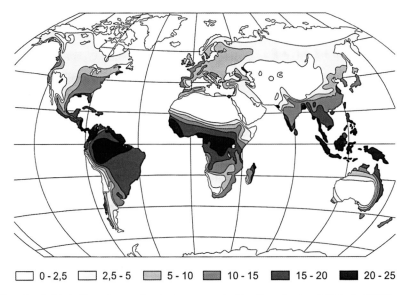

☐ 0 - 2,5   ☐ 2,5 - 5   ▨ 5 - 10   ▨ 10 - 15   ▨ 15 - 20   ■ 20 - 25

Fig. 4.2.   Areal distribution of annual net primary production (tons/hectare) (Schultz 2005).

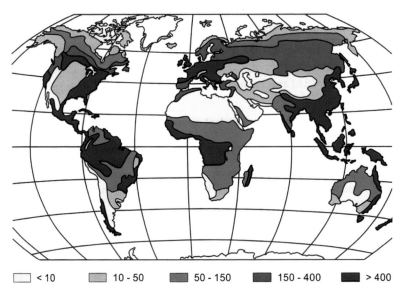

☐ < 10   ▨ 10 - 50   ▨ 50 - 150   ▨ 150 - 400   ■ > 400

Fig. 4.3.   Global distribution of surface and subsurface biomass (tons dry matter/hectare) (Schultz 2005).

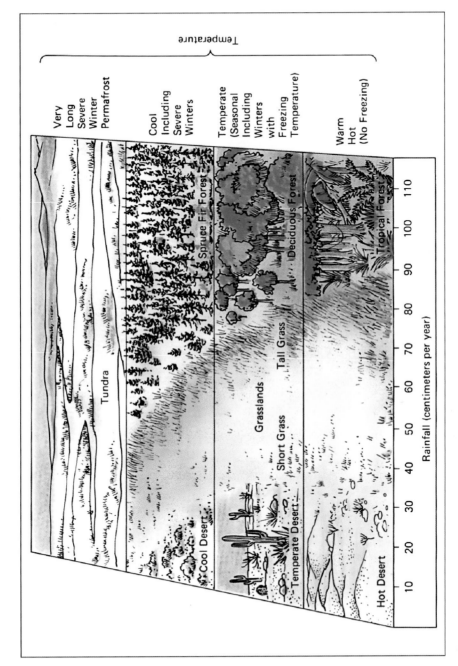

Fig. 4.4. Pictorial representation of two major factors shaping the world's biomes: rainfall and temperature (Nebel 1987).

soil types. Accordingly, the classification's principal soil orders are based both on the genetic modes of soil formation and on the recognizable characteristics of each soil type, including the presence or absence of indicative features such as diagnostic soil horizons. The orders we shall list below represent the highest category in a taxonomic hierarchy that includes suborders, great groups, families, and series. The two schemes draw on earlier efforts and are similar in fundamental principles, although they differ in several practical details, including the names applied to the various soil categories.

The FAO scheme was published in connection with the FAO/UNESCO *Soil Map of the World*. It retains some of the classical Russian names, while introducing many new ones for soil categories. This classification scheme includes 26 classes, most of which are roughly equivalent to the suborders and great soil groups of the USDA scheme. The latter recognizes ten major orders that are divided into suborders and further subdivided into great soil groups, subgroups, families, and series. The orders are characterized by diagnostic horizons, while lower levels in the classification hierarchy are based on such definable criteria as soil moisture and temperature regimes as well as on characteristic horizons.

Different countries and regions around the world have developed their own soil classification schemes that incorporate some of the same general principles as the comprehensive global schemes while including specific local features in each case. In our presentation, we mainly follow the US system.

Soils of Polar Regions (Tundras)

Soils of polar regions form in typically treeless plains in the arctic regions of Northern Europe, Northern Asia (Siberia), and Northern North America (Canada), where the land is either perpetually or intermittently frozen. Ice lenses often occur at shallow depths, and the soil may show evidence of physical churning *(cryoturbation)* due to the alternating cycles of freezing and thawing of water.

The vegetation consists of sphagnum moss and lichens, with a brief seasonal growth of grass. There may also be a sparse growth of shrubs. The soils tend to be shallow, acidic, to accumulate organic matter, and to be waterlogged during thaw periods. In places, such soils form *puffs* (or *blisters*) and depressions due to the uneven processes of frost heave. When thawed, such soils tend to emit methane *(marsh gas)*.

These soils are termed *gelisols* (from the Latin word *gelidus*, meaning "frost") in the US taxonomy system, and *cryosols* (from the Greek word *kruos*, "icy cold") in the FAO-UNESCO system.

Soils of Humid Temperate Regions

Soils of humid temperate regions occur typically in areas with a natural vegetation of deciduous or coniferous forests, extending from the subarctic zone to the Mediterranean region of Europe and stretching in a band between parallels 40 and 50 degrees latitude in Asia. They are also prevalent in Eastern North America. The soils often exhibit an A horizon consisting of a thin surface layer (A1) darkened by organic matter and underlain by a bleached layer

(A2) that resembles ash in appearance. Hence, these soils were called *podzols* (from the Russian words *pod* ["earth" or "soil"] and *zola* ["ash"]), in reference to the bleached light-grayish color of the A2 horizon. The process of *podzolization* has been linked to the influence of organic acids that leach downward from the organic-matter accumulation in the A1 horizon. In turn, the A2 horizon is underlain by a silicate-clay enriched B horizon.

Some podzolic soils tend to form hardened *pans* in the B horizon, due to the accumulation of iron oxides, clay, and leached components of organic matter. These soils are widely cropped. When subjected to erosion, they may lose the humus-containing A1 horizon and hence require the use of fertilizers to restore fertility, as well as liming to correct the soil's inherent acidity.

In the U.S. taxonomy system, a distinction is made between soils formed under deciduous forests, termed *alfisols* (presumably from the presence of aluminum and iron in the B horizon), and those formed under coniferous forests, called *spodosols* (from the Greek *spodos*, meaning "wood ash"). The latter tend to be strongly leached and acidic; hence, they are generally of lower inherent fertility.

Soils of Semiarid to Subhumid Plains (Steppes or Prairies)

Soils of semiarid to subhumid plains are typically grassland soils, with a natural vegetation of deeply rooted perennial grasses and shrubs that accumulate organic matter over time and to a considerable depth. The humus-enriched A-horizon is well aggregated and typically dark; hence, these soils were originally termed *chernozems* (Russian for "black earths"). Such soils occur in a belt running through parts of Southeast Europe (Hungary, Southern Poland [Galicia], and much of the Ukraine), as well as parts of central Asia (as far east as northern China). In North America, chernozems extend through the so-called "Wheat Belt" and "Corn Belt" of the Great Plains (from North Dakota to Kansas, and eastwards to parts of Minnesota, Iowa, Missouri, and Illinois). In South America, they occur in parts of Argentina's Pampas.

These soils, noted for their friable and stable structure and their legendary fertility, are termed *mollisols* (from the Latin root *mollis*, meaning "soft") in the US soil taxonomy system. In many areas, they are associated with a particular type of parent material called *loess*.[2]

Soils of Deserts and Semideserts

The category of soils of deserts and semideserts pertains to the soils of the warm lower latitudes, including North Africa, Southwestern Asia, Southwestern North America, Western South America, and parts of Australia. In addition, it includes some soils of arid inland regions in the cool middle latitudes of Central

---

[2] *Loess* (from a German root implying "loose") is typically composed of calcareous, buff-colored, fine silt, mainly of aeolian (wind-deposited) origin. When cut by a stream, loess deposits tend to form vertical walls. The most extensive areas of loess occur in Northern China in the basin of the Yellow River, which owes its name to the color of the silt it carries in suspension.

Asia and Mongolia, as well as the relatively dry parts of the Northern US and Canada, and parts of Patagonia in South America. In all of these regions, the limited rainfall produces only weak profile development. The main weathering processes are physical, and there is incomplete leaching of soluble constituents from the profile. The soil is dry much of the time, and it supports only sparse vegetation with incomplete surface cover.

The soils of arid regions are generally pale in color and contain little organic matter. They tend to retain the moderately soluble salts of calcium and magnesium. Where calcium carbonate accumulates and hardens in the subsoil, it may form an *indurated* (cemented) layer known as *caliche*. In other cases, calcium sulfate (gypsum) may crystallize at some depth. In extreme cases of aridity, especially in areas of impeded drainage, even the most readily soluble salts of sodium and potassium precipitate in the soil. Such soils are typically neutral to alkaline in reaction, but may be too saline for most crop plants and may allow only *halophytes* (specialized salt-loving plants) to survive.

Soils with a high content of sodium chloride were called *solonchak* by Russian pedologists or "white alkali" —from the color of salt crystals—by early American pedologists. Those with a predominance of sodium carbonate were called *solonetz* or "black alkali"—from the color of sodium-dispersed organic matter and the strongly alkaline reaction. Soils of semideserts were referred to as *serozems* in the old Russian system. In the modern USDA taxonomy system, such soils are termed *aridisols* (from Latin *aridus*, "dry"). In the FAO-UNESCO scheme, they are termed *calcisols, durisols, or gypsisols*, depending on what is regarded as the most prominent feature of the particular soil profile in the arid environment in which it evolved.

Soils of Humid Tropical Regions

In the humid tropics, the minerals in the parent rock generally undergo strong chemical weathering. The tendency is to hydrolyze and leach away the silica and to accumulate iron and aluminum oxides. As a result, the soils are typically colored red, the hue of haematite. The surface zone tends to be somewhat darker, but as a rule these soils do not exhibit distinct horizons. Chunks or blocks excavated from such soils and dried in the sun may harden to form bricks; hence, these soils have been called *laterites*, from the Latin word *later*, meaning "brick". Such soils are common in equatorial South America, Central Africa, Southeast Asia, Northern Australia, and tropical islands of the Pacific Ocean. In the US taxonomy system, they are termed *oxisols*—due to their accumulation of iron and aluminum oxides—whereas in the FAO-UNESCO system, they are named *ferralsols* or *plinthosols*. Particularly strongly leached and acidic tropical soils are termed *ultisols* (from the Latin *ultimus*, suggesting an "ultimate" or extreme condition of leaching). The more-or-less equivalent soils in the FAO-UNESCO system are termed *acrisols, alisols, or nitisols*.

Despite the often luxuriant natural vegetation, the soils of the humid tropics do not accumulate much organic matter, due to the rapid decomposition that occurs in the prevailing warm and moist conditions. When initially cultivated, the soils may be fertile for a few seasons but lose their fertility after the original organic matter reserve has been depleted. Hence, generations of indigenous farmers in Africa and Southeast Asia have practiced "shifting cultivation" (based on clearing and cultivating forested plots for several seasons, then allowing the forest to regenerate for two or three decades before returning to cultivate the same plots). Permanent cultivation requires considerable fertilization, liming to neutralize the soil's acidity, and stringent measures of conservation to prevent soil erosion.

Soils of Wetland Environments

Areas that are poorly drained tend to develop *hydromorphic* soils (called *glei* in the classical literature of pedology). They exhibit signs of chemical reduction rather than oxidation, including mottling (spots of reduced iron compounds, grayish rather than red) and occasional specks of elemental manganese. In such soils, the rate of accumulation of organic matter tends to exceed the rate of its decomposition.

Soils of marshes, moors, or bogs are typically rich in partly decomposed organic matter (peat), the anaerobic decomposition of which may emit methane ("marsh gas"). At saturation, these soils swell. Hence, when they are drained and aerated, the organic matter oxidizes and decomposes rapidly, and may even burn spontaneously. Drainage may also result in uneven shrinkage (subsidence) and the formation of hummocks. Such soils are termed *histosols* (from Greek *histos*, meaning "tissue") in both the US and FAO-UNESCO systems.

BOX 4.1 Peat, Mire, Moor, Marsh, Bog, Fen, Wetland and Weird Fire

Peat is an organic material formed in permanently saturated, anoxic, chemically reduced conditions. Locations where such conditions prevail are variously known as mires, moors, bogs, fens, or -- more generally -- marshes or wetlands. In places, such soils tend to bulge out progressively as they accumulate residues of plants such as sphagnums (belonging to the phytosociological classes of *Oxycocco-Sphagnetea* and *Vaccinio-Piceetea*) and the soil is generally a histosol. In places, the saturated areas surrounding raised bogs are called fens. Areas intermediate between bogs and fens are called transitional mires. However, the distinctions among the different types of hydric soils are unclear.

The anaerobic decomposition of organic matter in waterlogged soils typically releases methane (CH_4), which is therefore called "marsh gas." This gas forms bubbles in marsh sediments, and it may diffuse directly (or via hydrophytes) into the open air and thus contribute to the atmospheric greenhouse effect. In places, the emitted methane has been reported to ignite spontaneously, giving rise to a phenomenon called "will o' the wisp" or "ignis fatuus" ("delusional fire"), the flickering lights that are the source of many fantasies and legends. As pointed out by Gobat et al. (2004), however, methane does not ignite by itself and may only catch fire when mixed with another gas emitted from marshes, namely diphosphane (P_2H_4), which is formed by the chemical reduction of phosphates in a permanently saturated soil.

Soils of Swelling Clay in Wetting-Drying Conditions

Soils that contain a high percentage of expansive clay (such as the mineral montmorillonite) and are subject to alternating cycles of wetting and drying tend to swell and shrink markedly. When wet, they are sticky and miry. As they dry, they shrink and form deep cracks. The alternating movement of the soil forms shear planes within the profile that are called *slickensides*. Because of the tendency of the surface zone to break into small crumbs that dribble into the cracks, and because of the churning motion of the entire profile during repeated wetting-drying cycles, these soils mix themselves so that the profile shows little differentiation between horizons. Such soils were called *grumusols* in the older literature, but are now generally termed *vertisols* (from the Latin *verto*, "to turn").

Vertisols are common in the Deccan Plateau of the Indian Subcontinent, where they are called *black cotton soils*. Similar soils occur in the Gezira region of the Sudan, where the Blue Nile, with its load of clayey silt from the basaltic Highlands of Western Ethiopia, and the While Nile, conveying suspended organic matter from the swamps of Central Africa, converge. Such soils are called *gilgai* In Australia, *tirs* in Morocco, *Vleigrond* in South Africa, *Sonsosuite* in Nicaragua, and *sha chiang* in China. Though difficult to till because they are sticky when wet and extremely hard when dry, such soils can be highly productive if properly managed. Owing to their strong tendency to swell and shrink, vertisols also present problems to civil engineers whenever they wish to build roads or base other structures on such structurally active soils.

Soils Formed of Volcanic Ash

Soils formed of volcanic ash occur in areas where volcanic ejecta had been carried downwind from active volcanoes and had formed deposits of sufficient depth for soil formation to take place over time. They are especially prevalent along the Pacific Rim. Such soils can be very fertile thanks to their content of nutrients but are difficult to till where they are loose or occur on steep slopes. These soils are termed *andisols* or *andosols* (from the Japanese *ando*, referring to a "black soil").

Poorly Developed (Immature) Soils

Immature soils are mineral soils that have undergone only weak or rudimentary development of a soil profile, either because of their young age or because of some environmental factor or combination of factors that inhibits profile differentiation. Such soils may be found in recent deposits of alluvium or on sloping areas where the rate of erosion prevents the *in situ* development of a fully formed soil profile. Soils that are almost constantly dry or wet also tend to be poorly formed. This diverse group of soils is called *entisols* (implying "recent") in the US taxonomy system but are differentiated into several categories in the FAO-UNESCO, including *arenosols* (from the Latin *harena*, meaning "sand"), *fluviosols* (from the Latin *fluvius*, implying "riverine"), *leptosols* (from the Greek *leptos*, meaning "fine"), or *regosols* (from the Greek *rhegos*, meaning "blanket").

Fig. 4.5. Pictorial illustrations of principal soil orders. (Natural Resources Conservation Service, US Department of Agriculture 2006).

Alfisols are in semiarid to moist areas.

These soils result from weathering processes that leach clay minerals and other constituents out of the surface layer and into the subsoil, where they can hold and supply moisture and nutrients to plants. They formed primarily under forest or mixed vegetative cover and are productive for most crops.

ALFISOLS MAKE UP ABOUT **10%** OF THE WORLD'S ICE-FREE LAND SURFACE.

Fig. 4.5. Cont'd

Spodosols formed from weathering processes that strip organic matter combined with aluminum (with or without iron) from the surface layer and deposit them in the subsoil. In undisturbed areas, a gray eluvial horizon that has the color of uncoated quartz overlies a reddish brown or black subsoil.

Spodosols commonly occur in areas of coarse-textured deposits under coniferous forests of humid regions. They tend to be acid and infertile.

SPODOSOLS MAKE UP ABOUT 4% OF THE WORLD'S ICE-FREE LAND SURFACE.

Fig. 4.5. Cont'd

Mollisols are soils that have a dark colored surface horizon relatively high in content of organic matter. The soils are base rich throughout and therefore are quite fertile.

Mollisols characteristically form under grass in climates that have a moderate to pronounced seasonal moisture deficit. They are extensive soils on the steppes of Europe, Asia, North America, and South America.

MOLLISOLS MAKE UP ABOUT 7% OF THE WORLD'S ICE-FREE LAND SURFACE.

Fig. 4.5. Cont'd

Aridisols are soils that are too dry for the growth of mesophytic plants. The lack of moisture greatly restricts the intensity of weathering processes and limits most soil development processes to the upper part of the soils. Aridisols often accumulate gypsum, salt, calcium carbonate, and other materials that are easily leached from soils in more humid environments.

Aridisols are common in the deserts of the world.

ARIDISOLS MAKE UP ABOUT 12% OF THE WORLD'S ICE-FREE LAND SURFACE.

Fig. 4.5. Cont'd

Oxisols are highly weathered soils of tropical and subtropical regions. They are dominated by low activity minerals, such as quartz, kaolinite, and iron oxides. They tend to have indistinct horizons.

Oxisols characteristically occur on land surfaces that have been stable for a long time. They have low natural fertility as well as a low capacity to retain additions of lime and fertilizer.

OXISOLS MAKE UP ABOUT 8% OF THE WORLD'S ICE-FREE LAND SURFACE.

Fig. 4.5. Cont'd

Ultisols are soils in humid areas. They formed from fairly intense weathering and leaching processes that result in a clay-enriched subsoil dominated by minerals, such as quartz, kaolinite, and iron oxides.

Ultisols are typically acid soils in which most nutrients are concentrated in the upper few inches. They have a moderately low capacity to retain additions of lime and fertilizer.

ULTISOLS MAKE UP ABOUT 8% OF THE WORLD'S ICE-FREE LAND SURFACE.

Fig. 4.5. Cont'd

Histosols have a high content of organic matter and no perma-frost. Most are saturated year round, but a few are freely drained. Histosols are commonly called bogs, moors, peats, or mucks.

Histosols form in decomposed plant remains that accumulate in water, forest litter, or moss faster than they decay. If these soils are drained and exposed to air, microbial decomposition is accelerated and the soils may subside dramatically.

HISTOSOLS MAKE UP ABOUT 1% OF THE WORLD'S ICE-FREE LAND SURFACE.

Fig. 4.5.   Cont'd

Vertisols have a high content of expanding clay minerals. They undergo pronounced changes in volume with changes in moisture. They have cracks that open and close periodically, and that show evidence of soil movement in the profile.

Because they swell when wet, vertisols transmit water very slowly and have undergone little leaching. They tend to be fairly high in natural fertility.

VERTISOLS MAKE UP ABOUT 2% OF THE WORLD'S ICE-FREE LAND SURFACE.

Fig. 4.5. Cont'd

Andisols form from weathering processes that generate minerals with little orderly crystalline structure. These minerals can result in an unusually high water- and nutrient-holding capacity.

As a group, Andisols tend to be highly productive soils. They include weakly weathered soils with much volcanic glass as well as more strongly weathered soils. They are common in cool areas with moderate to high precipitation, especially those areas associated with volcanic materials.

ANDISOLS MAKE UP ABOUT 1% OF THE WORLD'S ICE-FREE LAND SURFACE.

Fig. 4.5. Cont'd

Entisols are soils that show little or no evidence of pedogenic horizon development.

Entisols occur in areas of recently deposited parent materials or in areas where erosion or deposition rates are faster than the rate of soil development; such as dunes, steep slopes, and flood plains. They occur in many environments.

ENTISOLS MAKE UP ABOUT 16% OF THE WORLD'S ICE-FREE LAND SURFACE.

Fig. 4.5. Cont'd

Inceptisols are soils of semiarid to humid environments that generally exhibit only moderate degrees of soil weathering and development.

Inceptisols have a wide range in characteristics and occur in a wide variety of climates.

INCEPTISOLS MAKE UP ABOUT **17%** OF THE WORLD'S ICE-FREE LAND SURFACE.

Fig. 4.5.   Cont'd

A group of soils with a somewhat greater degree of profile differentiation is called *inceptisols* (from the Latin *inceptum*, meaning "beginning") in the US taxonomy system. Such soils reveal a weakly developed B horizon that differs from the overlying A horizon in color and structure. Inceptisols have been identified, for example, in the riverine floodplains of the Ganges and Brahmaputra Rivers in Bangladesh and India, as well as in other floodplains in Southeast Asia. Some partially developed soils in the Sahel region of West Africa have also been included in this category. Such soils have been categorized as *cambisols* (from the Latin *cambire*, implying "exchanged") or *umbrisols* ("shadowed") in the FAO-UNESCO classification scheme.

RELATIVE ABUNDANCE OF SOIL ORDERS

An overall inventory of principal soil orders indicates that, of the Earth's total land area, a fraction of some 13% is occupied by exposed rocks (as in mountain ranges) or by shifting sands that cannot, strictly speaking, be considered in the category of *soil*.

Aridisols, occurring in deserts and their fringes, occupy some 12% of the Earth's land area. The grassland mollisols, foremost among the earth's inherently productive soils, cover only about 7% of the land area. The temperate-zone's forest soils, called alfisols, account for about 10%, while their sister soils, the spodosols, are estimated to cover about 3% of the planet's land area.

The frigid gelisols occupy as much as 9% of the Earth's land area. Their opposite type, the tropical oxisols, account for some 8% of the continental surface, and the latter's close relatives, the ultisols, make up another 8%.

The organic soils known as histosols account for only 1% of the world's land area, the clayey vertisols cover some 2%, and the volcanic andisols about 1%. The incompletely formed soils known as entisols extend over as much as 16%, and the young inceptisols cover about 10% of the globe's land surface.

The historical – and possible future – importance of some of these soil orders, however, far exceeds their seemingly small relatively areas, as may become evident in subsequent sections of this book.

BOX 4.2 The Human Impact on Soils

The human impact on the environment in general, and on soils in particular, began early in the history of our species. By the use of fire to open up areas for hunting, and later by the eradication of native vegetation for grazing and farming on an ever-increasing scale, human activity has resulted in the progressive erosion of topsoil and the depletion of its organic matter. Soils of low organic matter content tend to lose their optimal aggregated structure as well as their store of nutrient elements essential for plant growth. Consequently, such soils require greater inputs of fertilizers, soil amendments, and -- in some cases -- more energy-consuming tillage.

The areal extent and the intensity of human effects on the terrestrial domain has increased greatly in the last few centuries. Globally, some 3.25 to 3.47 million hectares (constituting about 10% of the total land surface) have been converted from natural vegetation to human use. This has had the overall effect of reducing the global net primary production (NPP) by about 5% and of emitting some 182 to 199 Pg of carbon to the atmosphere. In all, those changes in land cover and land use are responsible not only for the loss of soil organic matter but also for some 33% of the increased concentration of CO_2 in the atmosphere (Houghton et al., 1998).

A great question regarding the coming decades is whether the increasing demand for agricultural products due to growing populations and rising living standards will necessarily require progressive conversion of more forests and grasslands to croplands and pastures, or whether the increased production can be achieved by intensified management of selected land areas on a sustainable basis so as to allow the reversion of marginal (degradation-prone) land to natural habitats. In other words, must human appropriation of land be expanded at the expense of the remaining natural habitats, or can less land be managed more efficiently?

At present, the trend seems to be toward greater intensification (including high-efficiency water, nutrient, and even microclimate control) in the developed countries (mostly in temperate zones), but continued appropriation of natural biomes, especially deforestation (with attendant damage to biodiversity), in tropical and subtropical zones. The consequences of these trends, whether they will continue or may be changed, will be very large, perhaps decisive, with regard to the future of humanity, and quite possibly to the future of this planet's biosphere as a whole.

Global Soil Regions

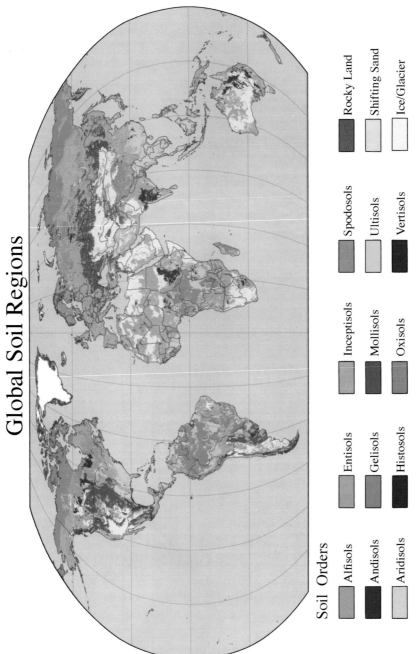

Fig. 4.6. Global distribution of principal soil orders. (Natural Resources Conservation Service, U.S. Department of Agriculture, 2006).

Soil Orders

Alfisols
Andisols
Aridisols

Entisols
Gelisols
Histosols

Inceptisols
Mollisols
Oxisols

Spodosols
Ultisols
Vertisols

Rocky Land
Shifting Sand
Ice/Glacier

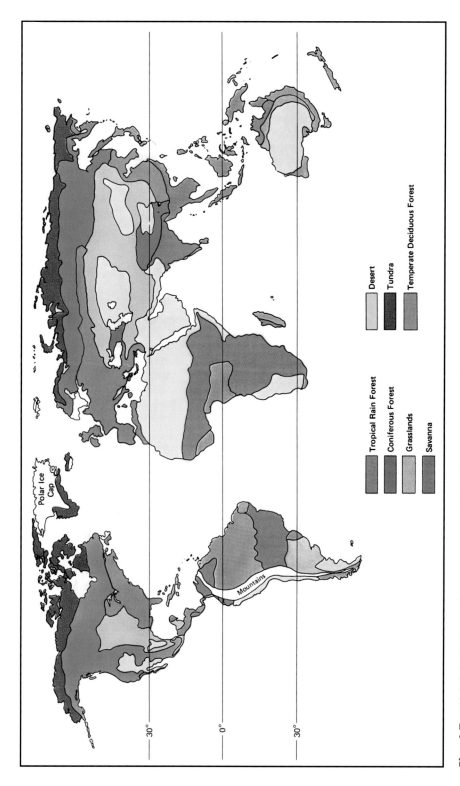

Fig. 4.7. Global distribution of the major vegetative biomes (note the correspondence with the distribution of soil orders). (Nebel, 1987).

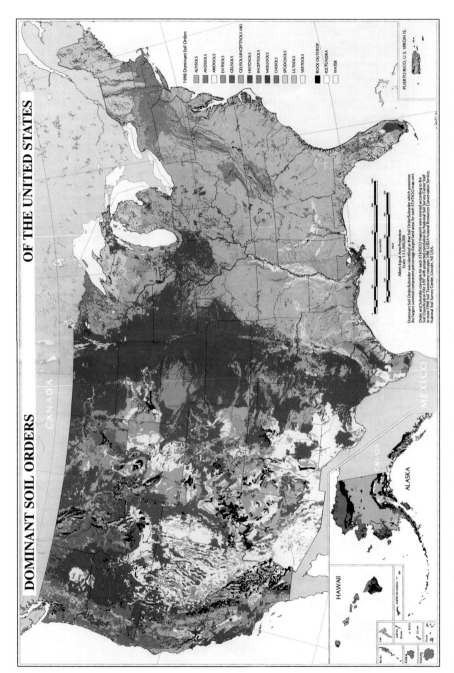

Fig. 4.8. Soils of the United States of America. (Natural Resources Conservation Service, U.S. Department of Agriculture).

5. SOIL PHYSICAL ATTRIBUTES

All these elaborately constructed forms, so different and dependent on each other, have all been produced by laws acting around us.
Charles Darwin (1809–1882)
The Origin of Species

SOIL TEXTURE

A body of soil typically includes three phases: (1) a *solid phase*, being an array of mineral particles of various shapes and sizes, often with amorphous compounds (e.g., organic matter or iron oxides) that are generally attached to the particles; (2) a *liquid phase*, consisting of water with varying mixtures and concentrations of solutes; and (3) a *gaseous phase*, constituting the air content of the soil. The solid phase forms the structural "skeleton" of the soil but occupies only a fraction (typically about 50%) of the soil's volume. Water and air in varying proportions permeate the labyrinthine interconnected spaces (called *pores*) that exist between the particles.

We can imagine a soil without air, or without water, or—in a vacuum—without both (as is the case with the "soil" found on the moon), but there could hardly be a soil without the solid phase. Hence, we shall focus on that phase first.

The array of particles in the soil may span several orders of magnitude. The largest particles are generally visible to the naked eye, whereas the smallest are colloidal and can be observed only with the aid of an electron microscope. It is customary to divide the particles into groups according to size and to characterize the solid phase as a whole in terms of the relative proportions of its

55

particle-size groups. Mineralogical analysis reveals that these groups generally differ from one another in mineral composition as well as in size. The attributes of the solid phase determine the nature and behavior of the soil as a whole: its internal geometry and porosity, its interactions with fluids and solutes, its thermal regime, and its mechanical attributes of compressibility and strength.

The term *soil texture* refers to the size range of particles in the soil. As such, this term carries both qualitative and quantitative connotations. Qualitatively, it describes the "feel" of the soil material, whether coarse and gritty or fine and smooth. An experienced soil classifier can feel, by kneading or rubbing the moistened soil, whether it is coarse-textured or fine-textured. In a more rigorously quantitative sense, however, soil texture defines the precisely measured distribution of particle sizes and the proportions by mass of the various size ranges of particles composing a given soil. As such, soil texture is an intrinsic, permanent attribute of the soil and the one most often used to characterize its physical makeup and behavior.

The importance of soil texture is that it provides an indication of how each soil might function in given circumstances. For example, a soil with large particles is also likely to have large pores, and consequently to be well aerated and very permeable to rainfall. After each rain, however, the water absorbed into such a soil is likely to drain away very rapidly. Hence, this soil will not retain much moisture for long during dry spells between successive rains. In contrast, a soil consisting of very fine particles and small pores may be less permeable to water but will retain more of the moisture it does absorb. The texture is also likely to have an effect on many other properties, including aeration, erodibility, mechanical stability, and thermal conductivity.

The dominant minerals in the both the sand and silt fractions are residues of the minerals in the original parent material; hence, they are known as primary minerals. Foremost is the weathering-resistant mineral quartz (SiO_2). Other minerals often present, though in smaller amounts, are mica, feldspars, zircon, haematite, and limonite. If the soil is not strongly leached, the sand and silt fractions may also contain fragments of calcite and dolomite.

Textural Fractions

The traditional method of characterizing soil texture is to divide the particles into three size-ranges, known as *textural fractions*; namely *sand, silt*, and *clay*. The procedure for making this separation is called *mechanical analysis*.

The conventional definition of *soil material* includes particles smaller than 2 mm in diameter. Larger mineral bodies are generally referred to as *gravel*, and still larger rock fragments are variously called *stones, cobbles*, or—if very large—*boulders*. While the presence of such bodies may affect the behavior of the soil, they are not in themselves considered soil material.

The largest particles generally recognized as soil material are categorized as *sand*, which is defined as particles ranging in diameter from 2000 micrometers to 50 (according to the USDA classification) or down to 20 micrometers (international classification). Sand grains usually consist of *primary minerals* (i.e., minerals present in the parent rock that have been fragmented but not altered chemically), such as quartz, feldspar, mica, and, occasionally, heavy minerals such as zircon, tourmaline, and hornblende. Some of the sand particles may

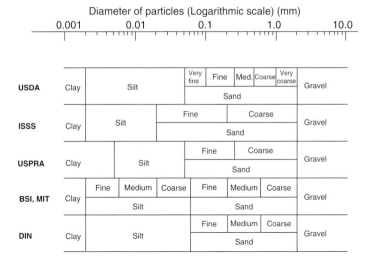

Fig. 5.1. Several schemes for classification of textural fractions.

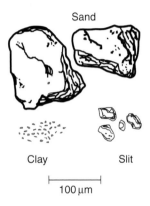

Fig. 5.2. Comparative sizes of sand, silt, and clay particles.

have smooth surfaces, but some may be jagged. That, together with their hardness, accounts for their abrasiveness.

In sharp contrast to sand, the smallest particles in the soil are classed as *clay*, defined as particles with a diameter smaller than 2 micrometers. Clay is the colloidal fraction of the soil. Its particles are typically plate-like or needle-like, and generally belong to a group of minerals known as *aluminosilicates*. These are *secondary minerals*, formed in the soil in the course of its evolution from the primary minerals contained in the original rock.

Because of its enormous surface area per unit mass, which may total many hundreds of square meters per gram, and its resulting physicochemical activity, clay is the fraction with the most influence on soil behavior. Clay particles adsorb water, forming hydration envelopes that cause the soil to well upon wetting and then to shrink markedly upon drying. Clay particles typically carry a negative electrostatic charge, and when hydrated, they attract cations from the aqueous solution present in the soil and repel anions. They then form an electrostatic double layer capable of exchanging cations with the surrounding solution. Another manifestation of surface activity is the heat that evolves when a dry clay is wetted, called the *heat of wetting*. A body of clay will typically exhibit plastic (moldable) behavior and become sticky when moist, but then cake up and crack to form hard fragments when desiccated.

The intermediate fraction between sand and clay is called *silt*. It consists of particles ranging in size between 2 and 20 micrometers (2 to 50 micrometers in the USDA system). Mineralogically as well as physically, silt particles generally resemble sand particles, but since they are smaller and have a greater surface area per unit mass and are often coated with strongly adherent clay, they may exhibit, to a limited degree, some of the physicochemical attributes of clay.

The relatively inert sand and silt fractions act as the soil's "skeleton", while the active clay, by analogy, acts as the soil's "flesh". It adsorbs water and solutes, swells and shrinks, becomes plastic and sticky when wet, and then becomes hard and brittle (often forming cracks) when dry. If treated with a monovalent cation such as sodium, clay disperses in water to form a suspension, but when treated with a divalent cation (e.g., calcium) it flocculates and settles out of suspension.

Together, all three fractions of the solid phase, as they are combined in various associations and configurations, constitute the *matrix* of the soil.

Textural Classes and Particle-Size Distribution

The textural designation of any particular soil, called its *textural class*, is determined on the basis of the mass ratios of the three fractions: sand, silt, and clay. The various classes are shown in Figures 5.3 and 5.4.

Note that a soil class called *loam* occupies a central location in the textural triangle. It refers to a soil that contains a "balanced" mix of coarse and fine particles with properties intermediate among those of a sand, a silt, and a clay. As such, loam is often considered the optimal soil texture for plant growth and agricultural cultivation. Its capacity to retain water and nutrients is superior to that of a sand, while its drainage, aeration, and mechanical properties are more favorable than those of a clay. However, under different environmental conditions and for different plant species, a sand (or sandy loam) or a clay (or clay loam) may be more suitable than a loam *per se*.

An alternative to dividing the array of soil particle into discrete size classes is to represent the particle-size distribution in a continuous graph, as shown in Figure 5.5. The ordinate of the graph indicates the percentage of the soil mass consisting of particles with diameters smaller than the diameter shown in the abscissa. The latter is drawn on a logarithmic scale to encompass several orders of magnitude of particle diameters. Soils with a smooth distribution curve are called *well-graded*, whereas soils having a preponderance of particles of one or several distinct sizes indicate a step-like distribution curve. This aspect of the particle size distribution can be expressed in terms of the so-called *uniformity index*, defined as the ratio of the diameter d60, which includes 60% of the particles, to the smaller diameter d10, which includes 10% of the particles.

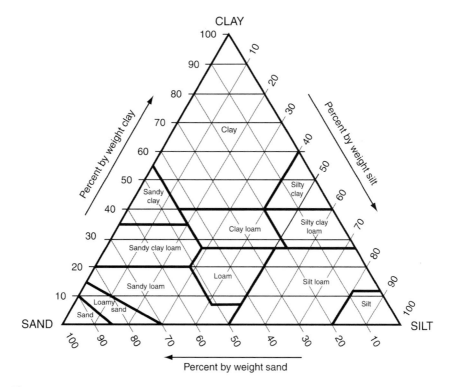

Fig. 5.3. The textural triangle, showing the mass percentages of clay (<0.002 mm), silt (0.002–0.05 mm), and sand (0.05–2.0 mm).

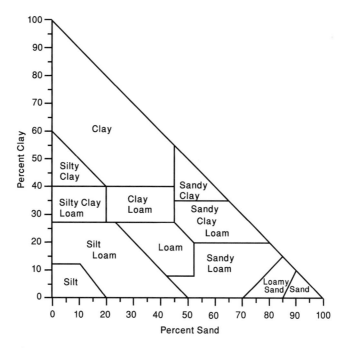

Fig. 5.4. Alternative representation of the textural triangle, showing the same textural classes.

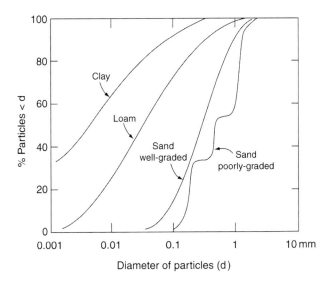

Fig. 5.5.   Particle size distribution curves for various types of soil.

Mechanical Analysis

Mechanical analysis is the traditional name for the procedure by which soil scientists determine the texture or particle-size distribution of a soil sample. The first step in this procedure is to disperse the soil sample in an aqueous suspension. The particles of the soil are often attached together in clusters called *aggregates*, and they must be separated and made discrete by removal of cementing agents (such as glue-like organic polymers). Moreover, inside the aggregates the tiny clay particles themselves are often clumped together in *flocs*, which also must be deflocculated for the purpose of measuring the soil texture.

Removal of organic matter is usually accomplished by oxidation with hydrogen peroxide, and dispersion of the clay flocs is done by treating the soil sample with a solution containing a chemical dispersing agent such as sodium metaphosphate, followed by mechanical agitation. Actual separation of the particles into size groups can be carried out by passing the suspension through graded sieves, down to a particle diameter of approximately 0.05 mm.

To separate and classify still finer particles, the method of sedimentation is usually used, based on measuring the relative settling velocity of particles of various sizes in an aqueous suspension. Naturally, the larger and heavier particles descend more rapidly, while the smaller and lighter ones descend more slowly and remain suspended longer. Any particle settling in a fluid encounters a frictional resistance proportional to the product of its radius and velocity, and to the viscosity of the fluid. Accordingly, the frictional resistance force F obeys the following equation, known as Stokes' law:

$$F = 6\pi\eta r u$$

where η is viscosity of the fluid, r is the radius of the particle, and u is its velocity.

One way to measure particle size distribution is to draw samples of known volume from a given depth in an initially stirred suspension at regular times after the stirring is stopped and sedimentation has begun. An alternative method is to use a hydrometer to measure the density of the suspension at a given depth as a function of time. With time, this density diminishes as the largest particles, and then progressively smaller ones, settle out of the region of the suspension being measured. This procedure has been standardized and a hydrometer specially calibrated for the purpose is available commercially.

SPECIFIC SURFACE

Many of the important phenomena and processes that occur in soils take place at the surfaces of particles. Hence, it is important to understand the nature and magnitude of those surfaces.

The *specific surface* of a soil material is generally defined as the total surface area of particles per unit mass, expressed in terms of square meters per gram. It can also be expressed as surface area per unit-volume of particles. Either way, it depends in the first place on the sizes of the particles—the smaller the particles, the larger their total surface area per unit mass. The specific surface also depends on the shapes of the particles. Flattened or elongated particles obviously expose a greater amount of surface per unit mass or volume than do equi-dimensional (i.e., cubical or spherical) particles.

Clay particles are typically platy (and sometimes needle-shaped); hence, they contribute even more to the overall specific surface area of a soil than is indicated by their small sizes alone. In addition to their external surfaces, certain types of clay crystals exhibit internal surface areas that form when the accordion-like crystal lattice of a clay mineral called *smectite* imbibes water. The specific surface of sand may be no more than 1 or 2 square meters per gram, whereas that of clay may be more than 100 times greater. So it is the content and mineral composition of the clay fraction that largely determines the specific surface of a soil.

The property of specific surface is a fundamental and intrinsic attribute of soils that has been found to correlate with important phenomena such as the soil's cation exchange capacity, adsorption and release of various chemicals, swelling and shrinkage, retention of water, and such mechanical properties as plasticity, cohesion, and strength. Hence, it is a highly pertinent property, perhaps more indicative of soil behavior than the percentage of sand, silt, and clay. It is generally measured by determining the amount of a gas or liquid needed to form a monomolecular layer over the entire surface of a soil sample in a process of adsorption. The standard method is to use an inert gas such as nitrogen, but water vapor as well as various organic liquids (e.g., glycerol and ethylene glycol) can also be used.

MINERAL COMPOSITION

The dominant minerals in both the sand and silt fractions are residues of the minerals in the original parent material; hence, they are known as *primary minerals*. Foremost is the weathering-resistant mineral quartz (SiO_2). Other minerals often present, though in smaller amounts, are mica, feldspars, zircon, haematite, and limonite. If the soil is not strongly leached, the sand and silt fractions may also contain fragments of calcite and dolomite.

The clay fraction differs fundamentally from the sand and silt fractions, not only in grain size but generally in mineralogy as well. It typically contains a class of minerals that are products of chemical weathering and re-precipitation, known as *secondary minerals* or *clay minerals*. They consist largely of alumino-silicates and hydrated oxides. The most prevalent of the clay minerals are crystal-line, though some clay minerals are amorphous; e.g., allophane and imogolite. The crystal-structured aluminosilicates are of three main types: (a) minerals with 1:1 alternating sheets of alumina and silica, such as in kaolinite and halloy-site; (b) minerals with 2:1 alternating layers of silica and alumina, such as illite, vermiculite, and smectite; and (c) minerals with 2:2 layers, such as chlorite.

The charges on the faces of the layered clay minerals result from the partial iso-morphous substitution of cations of lesser valence for those of higher valence. Such charges are "permanent" in the sense that they are independent of the pH of the enveloping soil solution. In addition, dissociated protons from the edges (rather than the faces) of clay crystals result in negative charges that are pH dependent.

The most loosely structured of the layered clay minerals is smectite, in which the crystal layers are so weakly bonded that water molecules can enter the interlayers and cause the mineral to swell, somewhat like an accordion. The tendency of this mineral to expand upon wetting and to shrink when drying may cause the soil to destabilize building foundations and to warp roadways.

THE NATURE OF CLAY

The term *clay* carries several connotations. In a qualitative sense, it suggests an earthy material that is soft and moldable when wet. In the more precise context of *soil texture*, it designates a range of particle sizes. Finally, in the mineralogical sense, it refers to a particular group of minerals, many of which occur in the textural clay fraction of the soil. Whereas sand and silt consist in the main of weathering-resistant primary minerals that were present in the original rock from which the soil was formed, clay includes secondary miner-als formed in the soil itself by chemical decomposition of the primary minerals and their recomposition into new ones.

The various clay minerals differ from one another in prevalence and proper-ties, and in the way they affect soil behavior. Rarely do any of these minerals occur in homogeneous deposits, and in the soil they generally appear in mix-tures, the composition of which depends in each case on the specific combina-tion of conditions that governed soil-forming processes.

The most prevalent minerals in the clay fraction of temperate region soils are the so-called aluminosilicates, whereas in moist tropical regions, hydrated oxides of iron and aluminum may predominate. The typical aluminosilicate clay minerals appear as laminated microcrystals, composed mainly of two basic structural units: a tetrahedron of four oxygen atoms surrounding a cen-tral cation, usually Si^{4+}, and an octahedron of six oxygen atoms or hydroxyls surrounding a somewhat larger cation of lesser valency, usually Al^{3+} or Mg^{2+}.

The tetrahedral are joined at their basal coroners by means of shared oxy-gen atoms in a hexagonal network that forms a flat sheet only 0.493 nm thick.

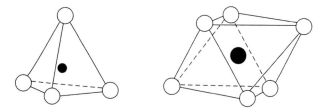

Fig. 5.6. The structural units of aluminosilicate clay minerals a tetrahedron of oxygen atoms surrounding a silicon ion (left) and an octahedron of oxygens or hydroxyls enclosing an aluminium ion.

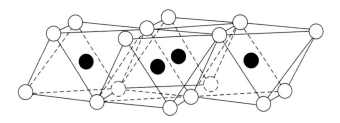

Fig. 5.7. Hexahedral network of tetrahedra forming a silica sheet.

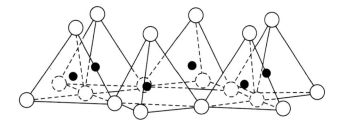

Fig. 5.8. Structural network of octahedra forming an alumina sheet.

The octahedral are similarly joined along their edges to form a triangular array. These sheets are about 0.505 nm thick.

The layered aluminosilicate clay minerals are of two main types, depending on the ratios of the tetrahedral to octahedral sheets, whether 1:1 or 2:1. In the first type–e.g. kaolinite–the unit cell is an octahedral sheet that is attached by the sharing of oxygens to a single tetrahedral sheet. In the second type, such as smectite (montmorillonite), the octahedral sheet is attached to two tetrahedral sheets, one on each side. In both cases, a clay particle is composed of multiple stacked unit cells of this sort.

The structure described is an idealized one. Typically, some substitutions of ions of approximately equal radii, called *isomorphous replacements*, take place during crystallization. In the tetrahedral sheets Al^{3+} may take the place of Si^{4+}, whereas in the octahedral layer Mg^{2+} may occasionally substitute for Al^{3+}. Consequently, internally unbalanced negative charges occur at different sites in the crystal lattice. Another source of unbalanced charge in clay crystals is the incomplete charge neutralization of terminal ions on lattice edges.

Clay minerals differ in their surface properties, including their charge densities (i.e., the number of electrostatic charges per unit area of surface) and

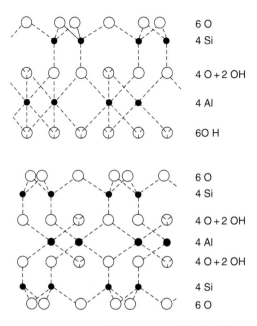

Fig. 5.9. Schematic representation of the structure of aluminosilicate minerals Kaolinite (top) and montmorillonite (bottom).

the cation exchange capacity, as well as the total surface areas per unit mass and the tendency to swell by allowing water and ions to enter into the spaces between crystal layers.

The negative electrostatic charges exhibited by most clay minerals in the hydrated state, particularly 2:1 minerals such as smectitie, are largely independent of pH. In some specific cases, however, the charges may be strongly dependent of pH. Such is the case with hydrous oxides of iron and aluminum (as well as with humus). Those oxides may even switch their charge from negative to positive due to adsorption of protons when the pH is lowered from neutral to strongly acidic. Some layered clay minerals may also manifest positive charges at the edges of their platey crystals, though the dominant charges on their planar faces are negative.

Table 5.1   Typical Properties of Prevalent Clay Minerals (Approximate Values)

Properties	Clay mineral				
	Kaolinite	**Illite**	**Montmorillonite**	**Chlorite**	**Allophane**
Planar Diameter (μm)	0.1–4	0.1–2	0.01–1	0.1–2	
Basic layer thickness (Å)	7.2	10	10	14	
Particle thickness (Å)	500	50–300	10–100	100–1000	
Specific surface (m²/g)	5–20	80–120	700–800	80	
Cation exchange capacity (mEq/100g)	3–15	15–40	80–100	20–40	40–70
Area per charge (Å²)	25	50	100	50	120

In highly weathered soil of the tropics, the clay fraction typically consists of hydrated oxides of iron, aluminum, and manganese. Among the prevalent minerals are goethite FeO(OH), haematite Fe_2O_3, gibbsite Al(OH), and birnessite (a manganese mineral of variable composition). The electrostatic charges on these poorly structured (amorphous) minerals are variable, generally depending on the pH of the ambient solution.

CATION EXCHANGE

The unbalanced charges of the clay must be compensated externally by the adsorption of ions (mostly cations) from the ambient aqueous solution. The adsorbed ions tend to concentrate near the external surfaces of the clay particles. When a clay particle is more or less dry, the neutralizing counter-ions are attached to its surface. Upon wetting, however, some of the ions dissociate from the surface and enter into solution. A hydrated particle of clay thus forms a *micelle*, in which the adsorbed ions are spatially separated, to a greater or lesser degree, from the negatively charged particle. As cations are attracted (i.e., positively adsorbed) to the particle, anions are generally repelled (i.e., negatively adsorbed) and thus relegated from the near proximity of the particle to the inter-micellar solution.

The adsorbed cations,(including Na^+, K^+, H^+, Mg^{2+}, Ca^{2+}, and Al^{3+}, hovering about the particle, are not an integral part of the lattice structure of the clay particle but are loosely associated with its surface and can be replaced, or exchanged, by other cations whenever the composition or concentration of the surrounding solution changes. Together, the charged clay surface and the diffuse swarm of cations form an *electrostatic double-layer* that responds dynamically to varying conditions. The *cation exchange* phenomenon is of great importance in soils, because it affects the retention and release of nutrients and other salts.

The concentration of solutes in the soil solution varies widely, depending on the water content itself, between saturation and air dryness, as well as on

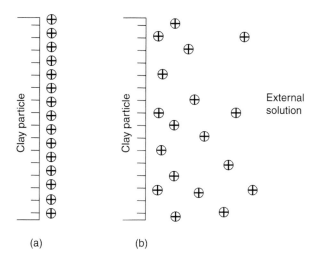

Fig. 5.10. Adsorbed cations associated with a negatively charged clay surface (a) when the surface is dry, and (b) when the surface is hydrated.

the general humidity or aridity of the climate. In general, the soil solution in humid regions, where the rainfall exceeds the potential evaporation, tends to be dilute, with a concentration typically of the order of 1 to 10 moles per cubic meter of solution. In contrast, the concentration of salts in the soil solution in arid regions, where the potential evaporation exceeds the amount of rainfall, may be as high as 50 to 100 moles per cubic meter. The species of ions most prevalent in the soil solution of non-acidic soils (with a pH value of 6 or above) are Ca^{2+}, Mg^{2+}, K^+, Na^+, NO_3^-, Cl^-, SO_4^{2-}, and HCO_3^-. In acidic soils, with a pH lower than 5, the soil solution often contains the ions Al^{3+}, $AlOH^{2+}$. The oxidation state of ions in the soil solution is strongly affected by soil aeration. Under anaerobic conditions, Mn^{++} and Fe^{++} are usually present.

The exchange of ions between the liquid phase (the soil solution with its ever-changing volume and solute composition) and the more-or-less stable solid phase depends greatly on the content and mineralogical composition of the clay fraction, as well as on the content and nature of the organic matter (humus) present in the soil. When Ca and Mg predominate, the soil is nearly neutral or slightly alkaline (having a pH value between 6.5 and 8). When Na and K predominate, the soil may become strongly alkaline (having a pH greater than 8.5). Finally, when H and Al ions predominate, the soil tends to become strongly acidic (having a pH below 5.5).

Ion exchange also affects the flocculation-dispersion processes of soil colloids and the tendency of soils to swell when imbibing water and to shrink during drying. These phenomena, in turn, affect soil structure. In principle, the swarm of adsorbed cations surrounding each particle acts as a sort of shield, repelling the cations of an adjacent particle, so the particles in effect remain separated from one another in the suspension. This condition, called *dispersion*, occurs especially when the cations are monovalent and the solution is dilute. However, when the monovalent cations are replaced by divalent or polyvalent cations, and when the ambient solution is concentrated, the ionic double-layer is compressed. Consequently, the particles can approach one another in the course of their random "Brownian" motion and "clump" together. This process of clumping is called *flocculation*. It causes the assemblages of particles, called *flocs*, to settle out of the suspension. Flocculation is the first step in the formation of soil aggregates, as we shall see in a subsequent section of this chapter.

Because of differences in their valences, radii, and hydration properties, different cations are adsorbed with different degrees of tenacity or preference, and hence are more readily or less readily exchangeable. Monovalent cations, attracted by only a single charge, can be replaced more easily than divalent or trivalent cations. The order of preference in exchange reactions is as follows:

$$Al^{3+} > Ca^{2+} > Mg^{2+} > NH_4^+ > H^+ > Na^+ > Li^+$$

An example of an exchange reaction is:

$$Na_2[clay] + Ca^{2+} \leftrightarrow Ca[clay] + 2Na^+$$

Such exchange reactions are rapid and reversible. The direction and extent of the exchange reactions vary, depending on the relative concentrations of the ions in the ambient solution and on their relative affinities to the adsorbing medium (clay or humus). In general, divalent ions are adsorbed more strongly than monovalent ions, but the relative affinity between clay and ions of different species also depends on the ionic concentrations. Also, as the composition and concentration of the soil solution change continuously during cycles of wetting and drying, as well as during the selective uptake of ions by plant roots, the composition of the soil's exchange complex is also likely to be highly labile.

The soil's exchange capacity is seldom adsorbed homogeneously with a single ionic species. Typically, the exchange capacity is taken up by several cations in varying proportions, all of which together constitute the soil's *exchange complex*. The total number of exchangeable cation charges, expressed in terms of chemical equivalents per unit mass of soil particles, is nearly constant for a given soil and practically independent of the species of cations present. It is thus considered to be an intrinsic property of the soil material and is generally called the *cation exchange capacity*, traditionally expressed in terms of milliequivalents of cations per 100 grams of soil. That capacity depends on the content and nature of the clay present in the soil. Sandy soils have a very low (sometimes negligible) exchange capacity, whereas soils containing much clay generally have a high exchange capacity. Of the various clay minerals, smectite has the highest and kaolinite has the lowest exchange capacity, with illite being intermediate.

In arid regions, calcium, magnesium, and sometimes sodium, tend to predominate in the exchange complex. In humid regions, where soils are highly leached and often acidic, hydrogen and aluminum ions play an important role. The presence of exchangeable hydrogen is usually referred to as *base unsaturation* of the exchange complex, and its measure in equivalent terms for a soil in an agricultural field is taken to be an indication of the amount of lime needed to neutralize the soil's acidity.

As cations are attracted to negatively charged sites on the faces of clay particles, anions are repelled. However, in some circumstances, anions may also be adsorbed; for example, on the edges of clay crystals as well as on the surfaces of amorphous masses of hydrated iron and aluminum. The adsorption of inorganic anions, however, is often blocked by the presence of organic anions. In some cases, mineral and organic anions may form composite complexes with cations, so the processes of adsorption, desorption, and exchange can be complicated indeed. Clay minerals in the soil also adsorb macro molecules of humus, which may act as "bridges" between particles, thus bonding them together.

Sorptive Properties

Clay minerals differ from one another in their surface properties. These differences pertain to the surface-density of the charges (i.e., the number of electrostatic charges per unit area of surface) and the cation exchange capacity, as well as to the total surface areas per unit mass and the tendency to swell and thus to allow water and ions to enter into the spaces between crystal layers.

As stated above, clay particles generally manifest a negative electrostatic charge. That charge may be largely independent of the pH, as it is in the case of 2:1 clay minerals such as smectite. In other cases the charge may be strongly dependent on pH. Such is the case with hydrous oxides of iron and aluminum. The latter oxides may even switch their charge from negative to positive due to the adsorption of protons when the pH is lowered from neutral to strongly acidic. Some layered clay minerals may also exhibit positive charges at the edges of their platy crystals, though the dominant charges on their "faces" are negative.

In the case of 2 :1 clay minerals, some of the cations are held in the narrow interlayer spaces, which restrict their diffusion to the external solution. Potassium ions, in particular, tend to lodge tightly between the silica sheets of illite crystals. In contrast, cations adsorbed on the particle's external surfaces and edges are more freely exchangeable whenever the composition and concentration of the ambient solution changes.

The ion exchange phenomenon, first discovered by soil scientists, has subsequently given rise to an industrial process of water purification. Both natural clay and artificial resins are used as active filtering media by which to purify contaminated water. When such water is percolated through the ion-exchanging medium, the potentially harmful components are adsorbed and retained, while the water emerges in purified form.

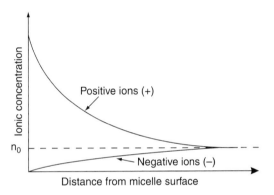

Fig. 5.11. Spatial distribution of cations and anions in the diffuse double-layer (NOTE: n_0 is the ionic concentration in the bulk solution).

Organic Colloids

Thus far in this chapter, we have discussed mainly the mineral colloids called clay. However, soils often contain an entirely different kind of colloidal matter known as *humus*. This generally dark-colored material, found mostly in the surface zone (the A horizon) can be defined as the more-or-less stable fraction of the soil organic matter remaining after the major portion of added plant and animal residues have decomposed. Note that this definition does not include undecomposed or partially decomposed organic matter, such as recent stubble or dead roots, which also exists in the soil.

Like clay, humus particles are negatively charged. During hydration, each particle of humus forms a micelle and acts like a giant, composite anion, capable of adsorbing various organic and inorganic constituents, including cations. The cation exchange capacity of humus is much greater per unit mass than that of clay, and humus may store, and slowly release, nutrients that promote plant growth.

Unlike most clay, however, humus is not crystalline but amorphous. Because it is composed mostly of carbon, oxygen, and hydrogen, its electrostatic charges are due not to isomorphous substitutions of cations but to the dissociation of carboxylic (–COOH) and phenolic (–◯–OH) groups. Since the cation exchange process depends on replacement of the hydrogen in these groups, it is pH dependent, with the cation exchange capacity generally increasing at higher pH values.

Humus is not a single compound, nor does it have the same composition in different locations. Rather, it is a complex mixture of numerous compounds, including lignoproteins, polysaccharides, polyuronides, and others too varied to list. Furthermore, the organic colloids of humus, although "more-or-less stable", are not nearly as stable as their mineral counterparts, and can be depleted gradually by oxidation and erosion, especially if the soil is excessively cultivated.

The content of humus in ordinary soils varies from 10% or more in the top layer of chernozems (mollisols), down to 0% in desert soils, and is in the order of 1–3% in many intermediate soils. The humus content generally diminishes in depth through the B horizon and becomes negligible at the bottom of the normal root zone. There are, however, special organic soils such as peat that can often be found in marshes and may contain over 50% organic matter, though not all of that would fit the accepted definition of humus.

The importance of humus exceeds its effect on cation adsorption and even its role in plant nutrition. Humus tends to coagulate in association with clay and serves as a cementing agent, binding and stabilizing assemblages of soil particles (called *aggregates*), thus improving soil structure (which is to be described in our next section). When virgin soils that are well drained are cleared of native vegetation and cultivated, a fraction of the humus originally present is gradually oxidized, along with most of the undecomposed organic residues. As a consequence, carbon dioxide is released into the atmosphere. Poorly drained soils, on the other hand, tend to emit methane, the gaseous product of unaerobic decomposition of organic matter. Both gases contribute to the atmospheric greenhouse effect and hence to global warming. Contrariwise, practices that retain or even augment the soil's organic matter content can help to mitigate global warming. We shall discuss those practices, now termed *carbon sequestration*, in much greater detail, in several subsequent sections of this book.

Soil Color

The color of a soil is greatly influenced by its organic matter content, which typically imparts a darkish hue to the topsoil, as well as by its mineral composition and degree of oxidation. In soils containing iron minerals, a reddish or yellowish hue indicates a well-aerated condition. A grayish or greenish, spotted ("mottled") discoloration, on the other hand, indicates poor aeration and reduction (rather than oxidation) of iron and other mineral elements. In some cases, where the soil is wetted and drained alternately, the exteriors of aggregates (clods) may be well oxidized and reddish, but the interiors of clods may be grayish due to locally anaerobic conditions. Compact and poorly drained (waterlogged) soils are likely to suffer restricted aeration, whereas porous, loose, and well drained soils tend to be well aerated.

SOIL STRUCTURE AND AGGREGATION

Thus far we have dealt with the components of the soil (e.g., sand, silt, clay, and humus) and their separate properties. To understand how the soil functions as a composite body, we must consider the manner in which those components are packed and held together to form a continuous spatial network. The arrangement or organization of the particles in the soil is called *soil structure*.

As soil particles differ in size, shape, and orientation, and can be variously associated and interlinked, the mass of them can form complex and irregular patterns that are very difficult to characterize in exact geometric terms. A further complication is the inherently unstable nature of soil structure and hence its inconstancy in time, as well as its nonuniformity in space. Soil structure is strongly influenced by changes in climate, biological activity, and soil management practices, and is vulnerable to disruptive forces of a mechanical or physicochemical nature.

However difficult it is to characterize precisely, soil structure is extremely important. It determines the volume fraction of the pore space (called *porosity*), as well as the shapes of the pores in the soil and the array of their sizes. Hence, soil structure affects the content and transmission of both air and water in the soil. Moreover, as soil structure influences the mechanical properties of the soil, it may also affect such disparate phenomena as germination, seedling emergence, root growth, tillage, overland traffic, and erosion. Agriculturists are usually interested in having the soil, at least in its surface zone, in a loose and porous and permeable condition. Engineers, who use the soil as building material or foundation, often desire a dense and rigid soil with minimal permeability to provide maximal stability and resistance to shearing forces. Ecologists, in contrast, are concerned with maintaining the soil's optimal structure so as to insure the health and sustainability of natural ecosystems.

In general, we recognize three broad categories of soil structure: *single grained*, *massive*, and *aggregated*. The first of these categories prevails when the particles are entirely unattached to one another, as it is in the case of a coarse granular soil (sand) or an unconsolidated deposit of desert dust. On the other extreme, when the soil is tightly packed in large cohesive blocks, as is sometimes the case with dried clay, the structure can be called *massive*. Between these two possibilities, we can recognize an intermediate condition (or range of conditions) in which the soil particles are associated in more-or-less stable packets of soil, known as *aggregates*.

The latter type of structure, called "aggregated", is generally the most desirable condition for plant growth, especially in the early stages of germination and seedling establishment. The formation and maintenance of stable aggregates is the essential feature of soil *tilth*, a term used by agronomists to describe that highly desirable, yet unfortunately elusive, physical condition in which the soil is an optimally soft, friable, and porous assemblage of stable aggregates, permitting free entry and movement of water and air, easy planting and cultivation; and unobstructed germination, emergence of seedlings, and growth of roots.

Soil aggregates are not characterized by any universally fixed size, nor are they necessarily stable. The visible aggregates, generally several millimeters to centimeters in diameter, are often called *peds* or *macroaggregates*. These are usually composites of smaller groupings, or *microaggregates*, which in turn are associations of the ultimate structural units, being the *flocs*, or clusters of clay particles. Those clusters attach themselves to, and sometimes engulf, the much larger primary particles of sand and silt. The internal organization of these various groupings can be studied by means of microscopy, including electron microscopy, using both scanning and transmission methods, the latter with thin sections of soil samples embedded in congealed plastic resins.

Shapes of Aggregates

The various shapes of aggregates observable in the field can be classified as follows:

1. *Platy*: Horizontally layered, thin and flat aggregates resembling wafers. Such structures occur, for example, in recently deposited clay soils.
2. *Prismatic* or *columnar*: Vertically oriented pillars, often six-sided, up to 15 cm in diameter. Such structures are common in the B horizon of clayey soils, particularly in arid regions. Where the tops are flat, these vertical aggregates are called *prismatic*, and where rounded, *columnar*.
3. *Blocky*: Cube-like blocks of soil, up to 10 cm in size, sometimes angular with well-defined planar faces. These structures occur most commonly in the upper part of the B horizon.
4. *Spherical*: Rounded aggregates, generally no larger than 2 cm in size, typically present in the A horizon. Such units are called *granules* or, where particularly porous, *crumbs*.

The shapes, sizes, and densities of aggregates generally vary among soils of differing characteristics, and even between layers within the profile of a single

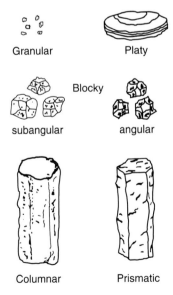

Granular   Platy

Blocky

subangular   angular

Columnar   Prismatic

Fig. 5.12.   Forms of soil aggregates.

soil. As the overburden pressure increases with depth, and the deeper layers do not experience such extreme fluctuations in moisture content as does the alternately saturated and desiccated surface layer, the decrease of swelling and shrinkage activities cause the deeper aggregates to be larger. A typical structural profile in a semiarid region consists of a granulated A horizon underlain by a prismatic B horizon, whereas in humid temperate regions, a granulated A horizon may occur with a platy or blocky B horizon. The number of variations found in nature is, however, very great.

Aggregate Stability

Any soil's state of aggregation is a time-variable property, as aggregates form, disintegrate, and re-form periodically. Soil structure may deteriorate as the soil is subjected to destructive forces resulting from slaking and erosion by water, trampling by animals, and tillage or compaction by heavy machinery. A prerequisite for aggregate stability is that the clay be flocculated. However, flocculation is a necessary but not sufficient condition for aggregation. An old truism in soil science is that "aggregation is flocculation plus". That "plus" is cementation, often caused by glue-like organic polymers or by iron oxides.

An important role in the evolution of aggregation is played by the extensive network of roots that permeate the soil and tend to enmesh soil aggregates. Roots exert pressures that compress aggregates and separate adjacent ones. Water uptake by roots causes differential dehydration, shrinkage, and the opening of numerous small cracks. Moreover, root exudations and the continual death and decay of roots promote microbial activity, which results in the production of humic cements. Some soil organisms form extensive adhesive filaments known as *mycelia* or *hyphae*, which also serve to promote aggregate stability. Prominent among the microbial products capable of binding soil aggregates are polysaccharides, hemicelluloses or uronides, levans, and numerous other natural polymers. Such flexible macromolecules form multiple bonds with several particles at once, and may form protective capsules around soil aggregates. Some organic products are inherently hydrophobic, so they may further promote aggregate stability by reducing wettability and swelling of clay. Because these various binding substances are transitory, being susceptible to further microbial decomposition, organic matter must be replenished continually if aggregate stability is to be maintained.

Active humus is accumulated and soil aggregates are stabilized most effectively under perennial sod-forming herbage. Annual cropping systems, on the other hand, hasten decomposition of humus and destruction of aggregates. The soil surface is especially vulnerable if exposed and desiccated in the absence of a protective cover. The foliage of close-growing vegetation and its residue protect topsoil aggregates against slaking by water, especially under raindrop impact. Some inorganic materials may also serve as cementing agents. Calcium carbonate, as well as iron and aluminum oxides, can impart stability to otherwise weak soil aggregates. The latter oxides are responsible for the remarkable stability of aggregates in some tropical soils with little organic matter.

Climatic factors, including freezing and thawing, as well as wetting and drying cycles that cause uneven stresses and strains in the soil and thus tend to compress and separate aggregates, can play an important role in the

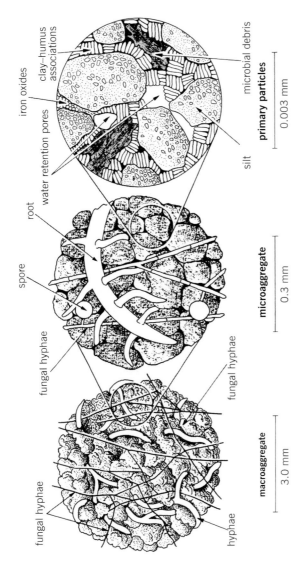

Fig. 5.13. The structural hierarchy of aggregates: Macroaggregates, microaggregates, and primary particles (Dubbin 2001).

formation of soil aggregates. The same factors, in different circumstances, can also contribute to the destruction of soil aggregates. The sudden wetting of a dry soil by, say, flooding or a downpour of rain, may trap and compress the air originally present inside the aggregates and literally cause them to explode—a process called *air-slaking*. Numerous small explosions of this sort can turn a well-aggregated soil surface into a puddled mass of mud that turns into a dense crust upon drying. The hammering impact of raindrops and the scouring action of surface runoff also contribute to aggregate breakdown. Finally, the actions of humans can have an extremely destructive effect on soil structure. Excessive grinding of the aggregates in the dry state by tillage or traffic may turn the soil surface into dust (vulnerable to deflation by wind, and—when the rains come—to erosion by water), just as excessive compression and shearing of the aggregates in the wet state by repeated trampling may render the soil surface tight and practically impermeable. For all the reasons described, the conservation of a well-aggregated structure is an essential task of soil management.

SOIL AIR AND GAS EXCHANGE

The fractional volume of air in the soil depends on the porosity (i.e., on the bulk density) and the volumetric water content. If the soil is aggregated, it is entirely possible that the macropores between the aggregates will be well aerated, while the micropores inside the aggregates will be anaerobic, owing to the restricted diffusion of air into the dense, water-retaining aggregates.

Where aeration is restricted and anaerobic conditions prevail over considerable periods of time, soil microorganisms tend to produce such gases as methane and nitrous oxide, both of which, incidentally, may contribute to global warming when emitted to the atmosphere. Other consequences of anaerobic conditions are the formation of chemically reduced forms of iron (Fe^{2+}), manganese (Mn^{2+}), sulfur (S^{2-}), ethylene, and organic acids, which may be toxic to the roots of some plant species.

The content and composition of the air phase in the soil is an important determinant of soil ecology. Roots of plants growing in the soil normally respire by absorbing oxygen and releasing carbon dioxide. Some specialized plants (called *hydrophytes*) are able to transfer oxygen from their above-ground parts (leaves and stems) to their roots internally, and thus can thrive in a saturated soil. However, most terrestrial plants (called *mesophytes*) require that the soil itself be aerated. In a well aerated soil, gaseous exchange takes place between the air phase of the soil and the external atmosphere at a rate sufficient to prevent a deficiency of oxygen and an excess of carbon dioxide from developing in the root zone. Aerobic microorganisms in the soil also respire and, under restricted aeration, might compete with the roots of higher plants for scarce oxygen.

Gases can move either in the air phase (i.e., in the pores that are drained of water, provided they are interconnected and open to the atmosphere) or in dissolved form through the water phase. The rate of transfer of gases in the air phase is generally much greater than in the water phase; hence, soil aeration depends largely on the volume fraction and continuity of the air-filled pores.

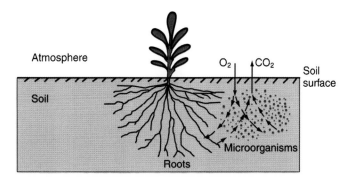

Fig. 5.14. Soil aeration is primarily a process of O_2 and CO_2 exchange between the air phase of the soil and the external atmosphere.

Impeded aeration resulting from restricted drainage and waterlogging, or from the mechanical compaction of the soil surface, can strongly inhibit the growth of mesophytes, especially of sensitive crop plants. Anaerobic conditions in the soil induce a series of chemical and biochemical reduction reactions (to be described more fully in a subsequent chapter). Among these reactions are *denitrification*, the process by which nitrate is reduced to nitrite and ultimately to elemental nitrogen; *manganese reduction* from the manganic to the manganous form, and, finally, to the elemental manganese; *iron reduction* from the ferric to the ferrous form; and *sulfate reduction* to hydrogen sulfide. Many organic compounds may result from the anaerobic decomposition of organic matter, including methane, ethylene, and various organic acids.

Under natural conditions, the volume ratios of the three constituent phases—solid particles, water, and air—are continually changing as the soil undergoes wetting and drying, swelling and shrinkage, aggregation and dispersion, loosening and compaction, etc. Since the two fluids, water and air, together occupy the pore space, which in favorable soils averages about 50% of the total soil volume but in compacted soils may be much less than that,, the respective fractional volumes of water and air are so related that an increase of one generally entails a decrease of the other. Thus,

$$a = f - w$$

where a is the volume fraction of air, f is the total porosity (the fractional volume of soil not occupied by solids), and w is the volume fraction of water. In the special case of swelling soils with a high content of expansive clay, however, the very change in water content w causes a change in the porosity f.

As long as a soil is well aerated, the composition of its air phase remains close to that of the external ("open") atmosphere, because the oxygen consumed in the soil by respiring roots and soil-borne microbes is readily replaced from the atmosphere while the carbon dioxide generated is readily vented to the atmosphere. Not so in a poorly aerated soil, where the CO_2 concentrations can sometimes be 10 or more times as great as normal atmospheric concentration (0.036%), with a corresponding decrease in the concentration of oxygen. In extreme cases, prolonged anaerobic conditions cause *anaerobiosis* (biological oxygen stress) and can result in the development of a chemical environment that induces reduction reactions such as denitrification, the evolution of such gases as hydrogen sulfide (H_2S), methane (CH_4), and ethylene (C_2H_4), and the reduction of mineral oxides such as those of iron and manganese.

The overall rate of respiration due to all biological activity in the soil depends on the nature and composition of the biotic community, as well as on the properties and conditions of the soil itself. Among the factors involved are the soil temperature regime, soil wetness, pH, and organic matter content and composition (whether fresh plant and animal residues or well-decayed humus), all of which influence the time-variable and space-variable activity of the organisms living in the soil.

Two main processes may operate to effect gaseous transport within the soil: *convection* and *diffusion*. Each of these processes can be formulated in terms of a linear rate law, stating that the rate of transport (the so-called *flux*) is proportional to the moving force. In the case of convection, also called *mass flow*, the moving force consists of a gradient of the total gas pressure, and it causes the entire mass of air to stream from a zone of higher pressure to one of lower pressure. In the case of diffusion, on the other hand, the moving force is a gradient of the partial pressure (or concentration) of any constituent member of the variable gas mixture that we call air. That gradient causes the molecules of the unevenly distributed constituents to migrate from a zone of higher concentration to one of lower concentration, even while the gas as a whole may remain isobaric and stationary.

Pressure differences between soil air and the outer atmosphere, inducing convective flow into or out of the soil, can be caused by barometric pressure changes, temperature gradients, and wind gusts. Additional phenomena affecting the pressure of soil air are the penetration of water during infiltration, causing displacement, and sometimes compression, of antecedent soil air; the fluctuation of a shallow water table, pushing air upward of drawing air downward; and the extraction of soil water by plant roots. Short-term changes in soil air pressure can also occur during tillage of compaction by machinery traveling over the surface.

The convective flow of air in the soil can be described by a simple equation:

$$q_v = -(k/\eta)\,\nabla P$$

where q_v is the volume convective flux of air (volume flowing through a unit cross-sectional area per unit time), k is the permeability of the air-filled pores network, η is the viscosity of the air, and ∇P is the three-dimensional gradient of soil air pressure. In one dimension, this equation takes the form

$$q_v = -(k/\eta)\,(dP/dx)$$

The process of diffusion can be formulated in terms of Fick's law:

$$q_d = -D(dc/dx)$$

where q_d is the diffusive flux of a gas (mass of a gas diffusing across a unit area per unit time), D is the diffusion coefficient, x is the distance, c is the concentration, x is distance, and dc/dx is the concentration gradient.

Considering the diffusive path in the air phase of the soil, we note that the diffusion coefficient in the soil D_s must be smaller than that in bulk air D_0 because of the limited fractional volume occupied by continuous air-filled pores and also because of the tortuous nature of these pores. Hence, we expect Ds to be some function of the air-filled porosity f_a partial volume of the air-filled porosity f_a. An oft-used approximation is the following:

$$D_s / D_0 = 0.66 f_a$$

In this formulation, 0.66 serves as a mean *tortuosity factor*, suggesting that the apparent (straight-line) path is about two-thirds of the length of the actual, typically tortuous, mean path of diffusion in the soil. This is admittedly a simplistic approximation. In reality, the magnitude of the tortuosity depends on the fractional volume of the air-filled pores, so it stands to reason that the tortuous path length should increase as the air-filled pore volume decreases.

In aggregated soils, the process is even more complex, since diffusion of gases in the relatively wide spaces between aggregates takes place much more rapidly than diffusion in the narrow pores inside the aggregates, which also tend to retain water long after the wide spaces have drained. Aggregates often have an anaerobic core that retains water and restricts oxidation.

Soil compaction, crusting, and waterlogging restrict the diffusion of oxygen to roots because of the blockage of some conducting pores and the increased tortuosity of others. In addition, some products of anaerobiosis, such as ethylene and carboxylic acids, restrict root growth.

BOX 5.1 Appearance of Poorly Aerated Soils

Poorly aerated soils often occur in humid regions characterized by high year-round rainfall. They may also occur in semihumid or even arid regions, in low-lying river valleys, deltas, and estuaries, as well as along seacoasts. They can generally be recognized by their waterlogged conditions, although occasionally a poorly aerated soil may appear to be dry on the surface. A pit dug into such a soil will often emit such gases as hydrogen sulfide, noted for its foul smell, and methane, which is highly flammable. Such gases result from the anaerobic decomposition of organic matter. Nitrogen is often reduced from its nitric state to a nitrous state, and it may even be reduced to the elemental N_2 gas that volatilizes to the atmosphere.

Prolonged anaerobic conditions result in the chemical reduction of mineral elements, particularly iron and manganese. Consequently, an anaerobic soil layer within the profile will acquire a bleached, grayish—occasionally even bluish or greenish—hue. A telltale sign of poor aeration is a condition known as mottling, characterized by the appearance of spotty discoloration, typically just above the water table. These spots generally indicate that some of the normally reddish ferric oxides have been reduced to ferrous oxides. In extreme cases, manganese is reduced from an oxide to the elemental state and occurs in the form of small blackish concretions.

Even apparently well-aerated soils may contain zones that are poorly aerated. In some cases, such zones occur as distinct layers, generally composed of tight clay. In other cases, they occur as isolated spots throughout the profile, as, for example, in the interiors of aggregates where water is retained for long periods. The outer portion of such an aggregate may be bright red and seem to be entirely oxidized, while the interior of the same aggregate may be mottled. Plant roots may grow all around the periphery of such aggregates but seldom penetrate into them.

6. SOIL-WATER STATICS

And with water we have made all living things.
The Koran, Sura 21:30

WATER CONTENT IN THE SOIL

The varying amount of water contained in a unit mass or volume of soil and the energy state of water in the soil are important factors affecting the growth of plants and all living organisms associated with the soil. Numerous other soil properties depend on water content. Among these are the mechanical properties of consistency, plasticity, strength, compactibility, penetrability, stickiness, and trafficability. In clayey soils, swelling and shrinking associated with additions or extractions of water change the porosity and density of the soil, as well as its pore-size distribution. Soil water content also governs the air content and gas diffusion in the soil, thus affecting respiration of roots, activity of microorganisms, and the chemical state of the soil; e.g., the oxidation-reduction potential.

The per-mass or per-volume fraction of water in the soil is expressed in terms of *soil wetness* (often called the *water content*):

$$w_m = M_w / M_s$$
$$v_w = V_w / V_t = V_w / (V_s + V_w + V_a)$$

Wherein w_m the mass wetness, is the dimensionless ratio of water mass M_w to dry soil mass, M_s; while v_w, the volume wetness, is the ratio of water volume V_w to total (bulk) soil volume V_t. The latter is equal to the sum of the volumes of solids (V_s), water (V_w), and air (V_a).

Both w_m and v_w are usually multiplied by 100 and reported as percentages by volume or mass. The two expressions can be related to each other by means

79

of the *bulk density* of the soil d_b, the ratio of mass of solids to the total volume including the pores, and the density of water d_w:

$$v_w = w_m (d_b / d_w)$$

The conversion is relatively simple for nonswelling soils in which the bulk density is constant regardless of wetness, but it can be difficult in the case of swelling soils for which the bulk density is a function of the mass wetness.

In many cases, it is useful to express the water content of a soil profile in terms of depth; that is, as the volume of water per unit area contained in a specified depth of soil. This indicates the equivalent depth that soil water would have if it were extracted and then ponded over the soil surface.

Measuring Water Content

The need to determine the amount of water contained in the soil arises frequently in many agronomic, ecological, and hydrological investigations. The traditional method of measuring mass wetness consists of removing a sample by augering into the soil and then determining its moist and dry weights. The moist weight is obtained by weighing the sample as it is at the time of sampling. The dry weight is obtained after drying the sample in an oven, generally at 105 degrees Celsius over a period of 24 hours. This so-called *gravimetric method* may seem simple, but it is in fact not trouble-free. The process of sampling, transporting, and repeatedly weighing samples entails practically inevitable errors. It is also laborious and time consuming.

An alternative approach is to embed a sensor in the soil in order to provide a continuous *in situ* sensing of soil moisture. Among such sensors are electrical resistance blocks, which generally contain a pair of electrodes embedded in a porous material, and which are inserted into the soil profile to various depths. The wires from such blocks are led to the soil surface, where they can be hooked to a monitor and read periodically or recorded continuously. Such blocks must be pre-calibrated against soil moisture content or suction.

A more sophisticated method of monitoring soil moisture *in situ* is the use of a neutron meter. The instrument consists of two main components: (1) a probe containing both a radioactive source that emits fast neutrons and a detector of slow neutrons that is lowered into a hollow access tube pre-inserted into the ground; and (2) a battery-powered scaler or rate meter to monitor the flux of the slow neutrons that are scattered and attenuated in the soil. The principle of the method is that the fast neutrons emitted radially into the soil collide elastically with nuclei of hydrogen atoms, and as they do so repeatedly, the neutrons lose much of their initial kinetic energy and are deflected every which way. The slow neutrons thus produced scatter randomly in the soil, forming a swarm of constant density around the probe. The density of the slow neutrons formed around the probe is nearly proportional to the concentration of hydrogen in the medium surrounding the probe, and therefore approximately proportional to the volume fraction of water present in the soil. The advantage of this method is that it permits repeated measurements in the same locations and depths. The disadvantages are the instrument's high cost and the potential health hazard that might be associated with exposure to neutron and gamma radiation. Precautions are therefore necessary when using this instrument repeatedly.

A relatively new method of measuring soil wetness, called *time-domain reflectometry (TDR)*, is based on the unusually high dielectric constant of water. This property, also called *relative permittivity* or *specific inductive capacity*, is a measure of the tendency of the molecules in a material to orient themselves in an electric force field. Because the molecules of water are inherently polar (or bipolar, having a positive side and a negative side due to the orientation of the hydrogen atoms relative to the oxygen in the H_2O molecule), the dielectric constant of water is relatively high, being about 81, whereas that of soil solids is of the order of 4 to 8, and that of air is about 1. Hence, when more water is present in the soil, the dielectric constant of the mixture increases. In practice, the method consists of inserting a pair of parallel metal rods, connected to signal receiver, into the soil. The rods serve as conductors while the soil between and around the rods serves as the dielectric medium. When a step voltage pulse is propagated along the parallel lines, the signal is reflected from the ends of those rods and returns to the TDR receiver. A device then measures the time between sending and receiving the reflected signal. For a fixed line length, the time interval relates inversely to the propagation velocity of the signal in the soil. Hence, that velocity diminishes as the amount of water present increases.

Additional approaches to the measurement of soil wetness include techniques based on gamma-ray absorption, on the dependence of soil thermal properties on water content, on the use of ultrasonic waves, radar waves, and microwaves, as well as on dielectric properties. Some of these and other methods have been applied to the remote sensing of land areas from aircraft or satellites.

ENERGY STATE OF SOIL WATER

Water in the soil is subject to several force fields that affect its potential energy state. Like other bodies in nature, soil water can carry energy in varying quantities and forms. Classical physics recognizes two main forms of energy: *kinetic* and *potential*. Because the movement of water in the soil is quite slow, its kinetic energy, which is proportional to the velocity squared, is generally considered to be negligible. However, the potential energy, which is due to position or internal condition, is of primary importance in determining the state and impelling the motion of water in the soil. The potential energy state of soil water also affects the vital functions of all organisms within the soil, including the extraction of water and nutrients by the roots of higher plants.

The potential energy per unit mass of water in the soil varies over a wide range. Differences in potential energy of water between one point and another give rise to the tendency of water to flow within the soil. The spontaneous tendency of all matter in nature is to move from where the potential energy is higher to where it is lower, so that each parcel of matter tends to equilibrate with its surroundings. Soil water obeys the same tendency toward elusive state known as *equilibrium*, definable as a condition of uniform potential energy throughout. The gradient of potential energy with distance is in fact the moving force causing flow.

Clearly, it is not the absolute amount of potential energy "contained" in the water that is important in itself, but the relative level of that energy in different regions within the soil. Just as an energy increment is the product of a force and a distance increment, the ratio of a potential energy increment to a distance increment can be viewed as constituting a force. Accordingly, the force acting on soil water is equal to the *negative potential gradient* ($-dF/dx$), which is the change of energy potential F with distance x. The negative sign indicates that the force acts in the direction of decreasing potential.

One way to consider the potential energy of soil water is to compare it to the potential energy of water in a standard state; i.e., the state of water in a hypothetical reservoir of pure "free" water at atmospheric pressure and at a reference elevation. An important component of the potential energy of water is its pressure. If a soil body at some elevation (say, on a hillslope) is saturated and its water is at a hydrostatic pressure greater than atmospheric pressure, the potential energy level of that water will then be considered positive relative to the standard reference state described. Water will then tend to move spontaneously from higher to lower pressure; i.e., from the soil into the atmosphere, as in the case of a spring discharging water from, for example, an aquifer. On the other hand, if the soil is moist but unsaturated, its water will no longer be free to flow out. On the contrary, the spontaneous tendency will be for the relatively dry soil to draw water from a reservoir at atmospheric pressure if placed in contact with it, much as a blotter draws ink or a dry napkin absorbs spilled coffee. An alternative statement of the same principle is that water tends to move spontaneously from a zone of higher to a zone of lower pressure potential.

The magnitude of the water potential at any point depends not only on hydrostatic pressure (whether positive or negative relative to some reference) but also on such additional physical factors as elevation, concentration of solutes, and temperature. The overall potential energy of soil water is higher at higher elevation, at higher temperature, as well as under higher pressure. However, it is lower where the concentration of solutes is higher. Accordingly, the total potential energy per unit mass of soil water, which we can call "the potential of soil water" for short, can be thought of as the sum of the contributions of those various factors, as follows:

$$F_t = F_g + F_p + F_o + F_t + \ldots$$

where F_t is the total potential, F_g the gravitational potential, F_p the pressure potential, F_o the osmotic potential, F_t the thermal potential, and the ellipsis signifies that additional terms are theoretically possible.

The gravitational potential arises from the fact that work must be expended to raise the elevation of a body; i.e., to lift it from a lower level to a higher one. That work is stored by the raised body in the form of potential energy, the amount of which per unit mass depends on the body's position relative to some reference level. So, a unit mass of water at the soil surface will have a higher gravitational potential than a unit mass of water at some depth within the profile. Note that the gravitational potential is independent of the chemical and pressure conditions of soil water, as it depends only on relative elevation.

The energy potential concept described is related fundamentally to the thermodynamic function known as "the specific differential Gibbs free energy". The "free energy" of water in an unsaturated soil is constrained (i.e., diminished) by the binding effect of adsorptive and capillary forces; that is to say, by the adsorption of water onto the surface of the grains and the attraction of water into narrow capillary pores. (The potential energy of water in the soil is also lowered by the presence of solutes; i.e., by the osmotic effect.) Owing to those forces, the effective hydrostatic pressure of water in an unsaturated soil is less than the atmospheric pressure of our "zero" reference state, which is a hypothetical reservoir of pure free water at a fixed elevation. Hence, water in an unsaturated soil is at an energy potential that we can consider "negative". Such a negative pressure has been called "tension" or "suction" in the soil physics literature. Still another name for that "negative" pressure potential of water in an unsaturated soil is *matric potential*. That name signifies that the subpressure of soil water in these circumstances results from the interactive forces between the water substance and the soil matrix.

In addition to the gravitational and pressure potential terms, we often need to consider the *osmotic potential*. The presence and concentration of solutes (generally salts) lower the potential energy of soil water. In particular, solutes lower the vapor pressure and reduce the vaporability of water. While this phenomenon may not affect liquid flow in the soil significantly (unless the concentration of salts is such as to significantly increase the viscosity and density of the soil solution), it definitely comes into play whenever an osmotic or diffusion barrier is present that transmits water more readily than salts. Thus, the osmotic effect is important in the uptake of soil water by plant roots, as well as in processes involving evaporation and vapor diffusion. There is a difference, however, between the osmotic potential and the other potential terms defined. Whereas the pressure and gravitational potentials refer to the soil solution (i.e., soil water along with its dissolved constituents), the osmotic potential applies to the water substance alone. Strictly speaking, therefore, F_o should not be simply added to F_m and F_g as if those terms

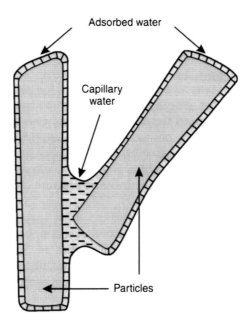

Fig. 6.1. Water in unsaturated soil is subject to capillarity and adsorption, which combine to produce a negative matric potential (suction).

were similarly applicable and mutually independent. This fundamental difference can only be ignored in practice as long as the soil solution is dilute enough and the solutes it contains do not produce a significant change in the matrix itself (e.g., by affecting the flocculation or dispersion of soil colloids) and therefore in the matric potential.

SOIL MOISTURE CHARACTERISTIC CURVE

The functional relation between a soil's water content and its suction is commonly represented in graphical form, called the "soil moisture characteristic curve". It is generally highly non-linear; in fact, it is not a single curve but a family of curves that differ from one another depending on their initial conditions. Those curves are bound between two envelope curves, known as the "wetting" and "drying" curves. Their displacement from one another is a measure of hysteresis.

Let us assume, as a starting point, that we have a soil sample that is completely saturated and at equilibrium with a body of free water at the same elevation; hence, soil water is at atmospheric pressure. In relative terms, we can say that the hydrostatic pressure of soil water is then zero. If a slight suction (i.e., a water pressure slightly subatmospheric) is applied to water in a saturated soil, no outflow may occur until, as suction is gradually increased, a critical value is reached at which the largest surface pore begins to empty and its water content is displaced by air. This critical suction is called the *air-entry suction*. Being the threshold of desaturation, air-entry suction is generally small in coarse-textured and in well-aggregated soils having large pores, but is relatively large in dense, poorly aggregated, fine-textured soils.

As suction is applied incrementally, the first pores to be emptied are the relatively large one that cannot retain water against the suction applied. A gradual increase in suction will result in the emptying of progressively smaller pores, until, at high suction values, only the very narrow pores retain water. Simultaneously, an increase in soil-water suction is associated with decreasing thickness of the hydration envelopes adsorbed to the soil particle surfaces. Increasing suction is thus associated with decreasing soil wetness. The amount of water remaining in the soil at equilibrium is thus a function of the matric suction. This function is usually measured experimentally and is represented graphically by a curve called the *soil moisture retention curve*, also known as the *soil moisture characteristic curve*.

The amount of water retained at low values of matric suction (say, between 0 and 100 kilopascals, or 0 to 1 bar) depends mainly on the capillary effect and the pore-size distribution. Hence, it is strongly affected by soil structure. At higher suctions, water retention is due increasingly to adsorption, so it is less influenced by the structure and more influenced by the texture and specific surface of soil. The greater the clay content, in general, the greater the water retention at any particular suction value, and the more gradual the slope of the curve. In a sandy soil, by contrast, most of the pores are relatively large, so once these large pores are emptied at a given suction, only a small amount of water remains.

Soil structure also influences the shape of the soil-moisture characteristic curve, primarily in the low suction range. A well-aggregated soil is characterized by a high water content at saturation and a relatively steep decrease of water content as suction increases. In comparison, a soil that has been compacted to the extent that its relatively large pores have been eliminated has a smaller volume of pores and exhibits a more gradual diminution of water content with increasing suction.

The slope of the soil moisture characteristic curve, which is the change of water content per unit change of matric potential, is generally termed the *differential* (or *specific*) *water capacity*. This term pertaining to moisture retention is analogous with the differential heat capacity, which is the change in the heat content of a body per unit change in the thermal potential (temperature). However, while the differential heat capacity is fairly constant with temperature for many materials, the differential water capacity in soils is strongly dependent of the matric potential. As such, it is an important property affecting soil moisture storage and availability to plants.

The relation between matric potential (suction) and soil wetness can be obtained in two ways: (1) in *desorption*, by starting with a saturated sample and applying increasing suction, in a step-wise manner, to gradually extract water while taking successive measurements of the remaining soil moisture as a function of the suction applied; and (2) in *sorption*, by gradually wetting an initially dry soil sample while reducing the suction incrementally. Each of these methods

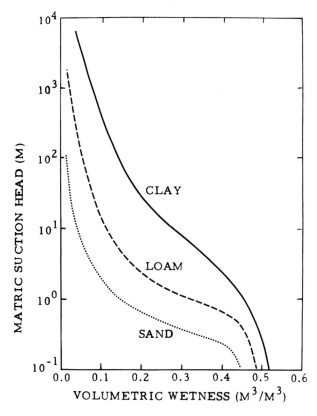

Fig. 6.2. Effect of texture on soil-water retention: Water content vs. suction for three textures.

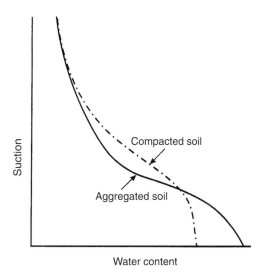

Fig. 6.3.   Comparative effect of aggregation vs.compaction on soil-water retention.

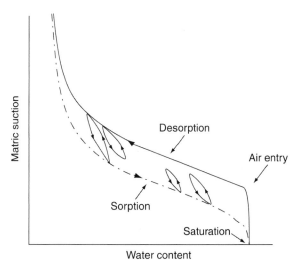

Fig. 6.4.   Hyssteresis in the suction vs. water content curves for sorption and desorption. (The intermediate loops represent transitions between the two main curves).

yields a continuous curve, but the two curves will generally not be identical. The equilibrium soil wetness at a given suction is greater in desorption than in sorption. This dependence of the equilibrium content and state of soil water upon the direction of the process leading up to it is called *hysteresis*.

The hysteresis effect may be attributed to several causes. The first is that the pores in the soil are not idealized smooth tubes but are irregularly shaped, with wide voids connected by narrow necks. In a system of such complex geometry, the processes of sorption and desorption do not follow the same pattern in reverse. A second reason for hysteresis is that the contact angle of water on the solid surfaces of grains tends to be zero in the case of desorption, as the water retreats over wet surfaces; that contact angle is greater than zero in the case of sorption, as the water advances over initially dry surfaces of the grains. Still another factor contributing to hysteresis is the encapsulation of air in "dead-end" pores, which reduces the water content of newly wetted soil. Finally, there are swelling, shrinkage, and aging phenomena that result in differential changes of soil structure, depending on the wetting and drying history of the sample. The gradual dissolution of air, or the release of dissolved air from soil water as a consequence of pressure and temperature changes, as well as microbial processes, can also have a differential effect on the suction-wetness relationship in wetting and drying systems.

MEASUREMENT OF SOIL WATER POTENTIAL

Several methods are available for the *in situ* measurement of the energy potential of soil moisture.

Perhaps the most common method is based on the use of an instrument called a *tensiometer*, designed to provide a continuous indication of the soil's matric suction *in situ*. The instrument consists of a porous cup, generally of ceramic material, connected through a tube to a manometer, with all parts filled with water. When the cup is embedded in the soil at the depth where the suction measurement is to be made, the bulk water inside the cup comes into hydraulic contact and tends to equilibrate with soil water through the pores in the ceramic walls. When initially placed in the soil, the water contained in the tensiometer is generally at atmospheric pressure. Soil water, being generally at subatmospheric pressure, exercises a suction, which draws out a certain amount of water from the rigid and airtight tensiometer. Consequently, the pressure inside the tensiometer falls below atmospheric pressure. The subpressure is indicated by a manometer, which may be a simple water-filled or mercury-filled U-tube, a vacuum gauge, or an electrical transducer.

A tensiometer left in the soil for a period of time tends to track the changes in the soil's matric tension, or suction. As soil moisture is depleted by drainage or by plant uptake, or as it is replenished by rainfall or irrigation, corresponding readings on the tensiometer gauge occur. Since the walls of the tensiometer's porous cup are permeable to both water and solutes, the solutes in the soil solution diffuse freely into the cup so that water in the tensiometer tends to assume the same solute composition and concentration as that of soil water. Therefore, the instrument does not indicate the osmotic suction of soil water, but only its matric suction. Another limitation of tensiometry is that its measurements are generally confined to the suction range of 0 to 1 atmosphere (about 1 bar, or 100 kilopascals). Tensiometers have long been used in guiding the timing of irrigation of field and orchard crops. Placing several tensiometers at different depths can indicate when the suction of soil water in the root zone approaches a critical value that requires irrigation to "refill" the "soil reservoir".

An alternative to the tensiometer is an instrument known as the *thermocouple psychrometer*. It is designed to indicate the relative humidity; that is, the ratio of the partial pressure of water vapor in the air phase of the soil to the equilibrium partial pressure of vapor in vapor-saturated air at the same temperature. The instrument generally measures the difference between the temperatures registered by a wet bulb and a dry bulb thermometer. The

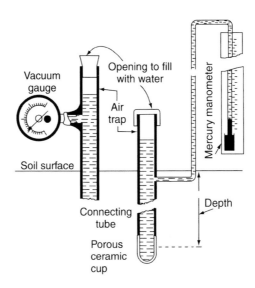

Fig. 6.5.   The essential parts of a tensiometer.

dry bulb thermometer indicates the temperature of a nonevaporating surface in thermal equilibrium with the ambient air. The wet bulb thermometer indicates the generally lower temperature of an evaporating surface, where latent heat is absorbed in proportion to the rate of evaporation. If the atmosphere has low relative humidity, its evaporative demand (and hence the evaporation rate) will be high, resulting in an appreciable depression of wet-bulb temperature below dry-bulb temperature. The relative humidity of air in equilibrium with a moist soil will depend on the temperature, as well as on the state of water in the soil. The constraining effects of adsorption, capillarity, and solutes act to reduce the evaporability of soil water relative to that of pure, free water at the same temperature. Hence, the relative humidity of an unsaturated soil's atmosphere will generally be under 100%, and the deficit to vapor saturation will depend on the soil-water potential, due to the combined effect of matric and osmotic potentials.

Recent decades have witnessed the development of miniaturized thermocouple psychrometers that make possible the measurement of soil-water potential *in situ*. A *thermocouple* is a double junction of two dissimilar metals. If the two junctions are subjected to different temperatures, they will generate a voltage difference. If, however, an electromotive force (emf) is applied between the junctions, a difference in temperature will develop. Depending on which way a direct current is applied, one junction can be heated while the other is cooled, and vice versa. The soil psychrometer consists of a fine wire thermocouple, one junction of which is equilibrated with the soil atmosphere by placing it inside a hollow porous cup embedded in the soil, while the other junction is kept in an insulated medium to buffer ambient temperature changes. During operation, an emf is applied so that the junction exposed to the soil atmosphere is cooled to a temperature below the dew point, at which point a droplet of water condenses on the junction, allowing it to become, in effect, a wet bulb thermometer.

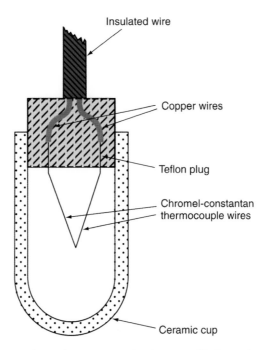

Insulated wire

Copper wires

Teflon plug

Chromel-constantan thermocouple wires

Ceramic cup

Fig. 6.6. Thermocouple psychrometer contained in an air-filled ceramic cup.

(This is the so-called *Peltier effect*.) The cooling is then stopped, and as the water from the droplet reevaporates, the junction attains a wet bulb temperature. Thus, the difference in temperature between the web-bulb and dry-bulb temperatures can be measured, and it indicates the soil-water potential.

Yet another method for measuring soil-water potential is based on the rate of heat dissipation in the soil. The rate of propagation of heat obviously depends on the thermal conductivity, which, in turn, is a measurable function of the soil moisture potential. In practice, the method consists of embedding both a heat source and a thermal sensor in a porous ceramic block, which is inserted into the soil to a desired depth. The temperature is measured just before and a fixed time after the application of a standard heat pulse, and the difference is indicative of the soil-moisture potential. The advantage of this method is that it applies beyond the limited range of a tensiometer. The main disadvantage, however, is that it requires very careful calibration and may be affected by hysteresis. The calibration can be done against the soil-water content as well as against the soil-water potential.

7. SOIL-WATER DYNAMICS

*Remember when discoursing about water
to adduce first experience and then reason*
Leonardo da Vinci, 1452–1519

WATER FLOW IN SATURATED SOIL

We first consider the movement of water in a saturated soil; i.e., when all the pores are filled with water to the exclusion of air. Even in this relatively simple case, the geometric pattern of flow is complex, as the pores themselves are irregularly shaped, tortuous, and intricately interconnected. Flow through the network of pores is limited by numerous constrictions, or "necks", with occasional "dead-end" spaces. The direction and velocity of flow varies drastically from point to point even along the same passage, and varies even more among different passages. Indeed, the actual geometry of flow is too complex to be described in microscopic detail. For this reason, flow through a porous medium such as a soil is generally described in terms of a macroscopic flow-velocity vector, which is the overall average of the microscopic velocities over the total volume considered. The detailed flow pattern is thus ignored, and the conducting body is treated as though it were a uniform medium, with the flow spread out over the entire cross-section, solid and pore space alike.

We next describe the quantitative relations connecting the rate of flow, the dimensions of the body, and the hydraulic conditions at the inflow and outflow boundaries. The figure below shows a horizontal column of soil through which a steady flow is occurring from left to right, from an upper to a lower reservoir, in each of which the water level is kept constant. Experience shows that the discharge rate Q, which is the volume V flowing through the column per unit time, is directly proportional to the cross-sectional area and to the

91

hydraulic head drop ΔH, and inversely proportional to the length L of the column:

$$Q = V/t = A\,\Delta H/L$$

The hydraulic head drop per unit distance ($\Delta H/L$) is called the *hydraulic gradient*, which is, in fact, the force driving the flow. The specific discharge, Q/A, namely the volume of water flowing across a unit area per unit time, is called the *flux density* (or simply the *flux*) and is indicated by q. Thus, the flux is proportional to the hydraulic gradient:

$$q = Q/A \propto \Delta H/L$$

The proportionality factor K is termed the *hydraulic conductivity*:

$$q = K\,\Delta H/L$$

This equation is known as *Darcy's Law*, after Henry Darcy, the French engineer who discovered it a century and a half ago in the course of his classic investigation of seepage through sand filters in the city of Dijon, France. A more generalized expression of this relationship is:

$$q = -K\,dH/dx$$

This generalized law states that the flow of a viscous liquid through a porous medium occurs in the direction of, and at a rate proportional to, the driving force (i.e., the hydraulic gradient) and is also proportional to the property of the medium called the hydraulic conductivity, a measure of the ability of the medium to transport water. The symbol q represents the volume of water passing through a unit cross-sectional area perpendicular to the flow direction per unit time. It is called the *flux density*, or simply the *flux*. The negative sign is

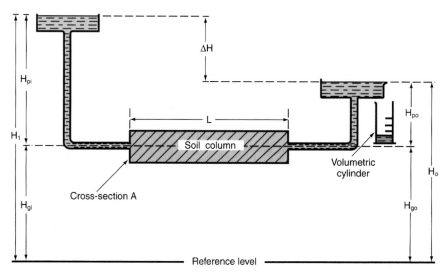

Fig. 7.1.   Flow in a horizontal saturated column.

due to the fact that flow occurs against the pressure gradient (i.e., from higher to lower pressure).

In the case of the horizontal column depicted above, the only driving force is the pressure-head gradient, dH/dx. In the case of a vertical column, however, there is another potential driving force, namely the gravitational-head gradient (depicted in the figure below), which in the case of downward flow equals unity.

Hydraulic Conductivity of Saturated Soil

The hydraulic conductivity, being the ratio of the flow rate (flux) to the potential gradient, can be expressed in different units. If the dimensions of the flux are volume per unit area per unit time (length per unit time), we note that the dimensions of hydraulic conductivity depend on those assigned to the potential gradient. The simplest way to express the potential gradient is by use of length, or head, units (though, strictly speaking, H is not a true length but a pressure equivalent in terms of a water column height $H = P/\rho g$). Therefore, the hydraulic head gradient H/L, being the ratio of length to length, is dimensionless. Accordingly, the dimensions of hydraulic conductivity are the same as those of flux, namely L/T.

In a saturated soil of stable structure, or in any rigid porous medium such as sandstone, the hydraulic conductivity is often assumed to be constant. Its order

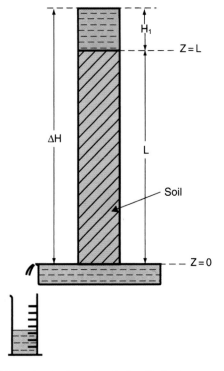

Fig. 7.2. Downward flow of water in a saturated vertical column.

of magnitude in a clay soil with typically narrow pores ranges from 10^{-6} to 10^{-9} m/sec. In a sandy soil with wider interparticle pores, the hydraulic conductivity may be of the order of 10^{-4} to 10^{-5} m/s.

To appreciate the practical significance of these values in more familiar terms, consider the hypothetical case of an unlined (earth-bottom) reservoir or pond. Assume, for the sake of simplicity, that the seepage from the reservoir occurs by gravity alone (i.e., no pressure gradient), at a rate equal to the hydraulic conductivity. A coarse sand with a K value of, say, 10^{-4} m/s would therefore lose water at the rate of nearly 9 m/day (there being 8.64×10^4 seconds per day). A loam soil with a K value of 10^{-6} m/s would lose "only" about 0.1 m/day (100 mm/day). Finally, in contrast, a bed of clay with a conductivity of 10^{-9} m/s would allow the seepage of no more than 1 mm/day, less than the generally expectable rate of evaporation. Clearly, then, the retention of water in earthen dams and the control of seepage from unlined canals can be greatly aided by packing the bottom with a bed of clay.

The hydraulic conductivity is affected by structure as well as by texture, being greater if the soil is highly porous or aggregated than if it is tightly compacted and dense. Hydraulic conductivity depends not only on total porosity but also—and primarily—on the sizes of the conducting pores. That is the reason why the conductivity of a bed of gravel or sand with large pores can be much greater than the conductivity of a clay soil with narrow pores, even though the total porosity of a clayey soil is generally greater than that of a sandy soil.

Contrary to the convenient assumption stated above that the hydraulic conductivity of a saturated soil remains constant, it is sometimes found to vary gradually over time. Because of chemical, physical, and biological processes, the K value may change due to ion exchange processes, varying concentration and composition of the permeating aqueous solution, flocculation and dispersion phenomena, gradual swelling of clay, migration and lodging of clay particles (possibly resulting in clogging of pores), as well as microbial activity (bacterial or fungal growth and its slimy or gaseous products). In practice, it is difficult to saturate a soil with water without trapping some air. Encapsulated air bubbles may block pore passages. Temperature changes affect the solubility of gases in water, and may thus cause the flowing water to dissolve or to release bubbles as well as to change their volume. Temperature also affects the viscosity (or fluidity) of water, which in turn affects the hydraulic conductivity.

WATER FLOW IN UNSATURATED SOIL

Most of the processes involving soil-water interactions in the field occur while the soil is in an unsaturated condition. These processes include the supply of moisture and nutrients to plant roots as well as the drainage of water and solutes beyond the root zone. Unsaturated flow often entails changes in the state and content of soil water, involving complex interrelations among variables such as soil wetness, suction, and conductivity.

In the preceding sections, we stated that the flow of water in the soil is driven by a hydraulic potential gradient, that it takes place in the direction of decreasing hydraulic potential, and that its rate is proportional to the potential gradient. These principles apply in unsaturated, as well as saturated, soils.

However, the nature of the moving force and the effective geometry of the conducting pores can be very different. Apart from the gravitational force, which is completely independent of soil water content, the primary moving force in a saturated soil is the gradient of a positive pressure potential. On the other hand, water in an unsaturated soil is subject to a sub-atmospheric pressure, or matric suction that is equivalent to a negative pressure potential. The gradient of this potential likewise constitutes a moving force.

Matric suction is due, as we have noted, to the physical affinity between water and the matrix of the soil. That affinity includes both the adsorption of water onto particle surfaces and the attraction of water into capillary pores. When suction is uniform along a horizontal body of soil, the soil is at equilibrium and there is no moving force. Not so when a suction gradient exists, in which case water will be drawn from a zone where the matric suction is lower to where it is higher. It will flow in the pores that are still water-filled and creep along the hydration films over the particles surfaces, in a tendency to equilibrate the potential. (The ideal state of equilibrium, like that of human happiness, may never be achieved in practice, but its natural pursuit is a universal rule!)

Hydraulic Conductivity of Unsaturated Soil

One of the most important differences between unsaturated and saturated flow pertains to the hydraulic conductivity. When a soil is saturated, all its pores are water-filled and conducting. The water phase is then continuous and the conductivity is maximal. When the soil desaturates, some pores become air-filled so that the conductive portion of the soil's volume diminishes. Furthermore, as suction develops, the first pores to empty are the largest ones, which are the most potentially conductive. At the same time, those large pores must be circumvented, so that with progressive desaturation, tortuosity increases, as does effective length of the flow path and hence the hydraulic resistance. In coarse-textured soils, water may be confined mainly to the capillary wedges at the contact points of the particles, thus forming separate and discontinuous pockets of water. In aggregated soils, too, the large interaggregate spaces that confer high conductivity at saturation become, when emptied, barriers to liquid flow from one aggregate to another. As high suctions occur, there may also be a change in the viscosity of the mainly adsorbed water, tending to further reduce the conductivity. (Viscosity is temperature-dependent as well.)

For these reasons, the transition from saturation to unsaturation generally entails a steep drop in hydraulic conductivity that may diminish by several orders of magnitude as suction increases from 0 to, say, 1 Mpa. At still higher suctions (i.e., lower wetness values), the conductivity may be so low that very steep suction gradients, or very long times, are required for any appreciable flow to occur at all.

The conductive properties of unsaturated soils depend greatly on their texture. Whereas at saturation a sandy soil conducts water more rapidly than a clayey soil, the opposite is often the case when unsaturated conditions prevail. While in a soil with large pores, these pores quickly empty and become nonconductive as suction develops, in a soil with small pores, many of the pores retain and conduct water even at appreciable suction so that the hydraulic

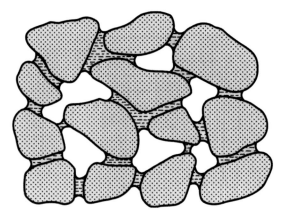

Fig. 7.3.   Water in an unsaturated coarse-textured soil.

conductivity does not decrease as steeply and may actually exceed that of a soil with large pores that is subjected to the same suction.

The following figure shows the dependence of conductivity on suction in soils of different texture. Note that K versus suction curves is usually drawn on a logarithmic scale, because K varies over several orders of magnitude within the suction range of general occurrence in the field.

The decline of hydraulic conductivity with increasing matric suction carries important implications regarding soil-water dynamics. It suggests that processes occurring in wet soil conditions are inherently faster than those in drier conditions. Thus, the process of infiltration, during which water moves into the soil profile through a saturated surface zone at maximal conductivity, is much more rapid than the process of evaporation, which typically involves the

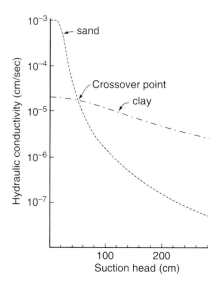

Fig. 7.4.   Dependence of hydraulic conductivity of a sand and a clay on suction.

transfer of water from the interior of the soil to the atmosphere through a drying surface zone of greatly diminished conductivity. A single large storm can therefore charge up an entire soil profile in just a few hours with an amount of water that subsequent evaporation can only extract very slowly over many weeks.

Flow Equation for Unsaturated Soil

Darcy's Law, originally conceived for flow in saturated porous media, has since been applied to unsaturated flow, with the provision that the conductivity is now a function of the matric suction; i.e., $K = K(\psi)$:

$$q = -K(\psi)\nabla H$$

where ∇H is the hydraulic head gradient, which may include both suction and gravitational components.

To obtain the general flow equation and account for transient, rather than steady, flow processes in which the soil medium may be gaining or losing water, we introduce the *continuity principle*, which suggests that the change of flux with distance must entail a change of water content with time:

$$\partial w / \partial t = -\nabla \cdot q$$

where the symbol ∇ represents the three-dimensional gradient in space (in our case, of the flux q). Note that the negative sign indicates that if the rate of flow accelerates through a given spatial domain, the water content must diminish in time within that domain, and vice versa.

The one-dimensional horizontal form of this equation is:

$$\frac{\partial w}{\partial t} = \frac{\partial}{\partial x}\left[K(\psi)\frac{\partial \psi}{\partial x}\right]$$

Where K is constant, this equation reduces to:

$$\frac{\partial w}{\partial t} = K\frac{\partial^2 H}{\partial x^2}$$

More typically in an unsaturated soil, the hydraulic conductivity K is itself a function of water content, w. The combination of Darcy's equation with the continuity equation, with the added proviso that the conductivity is a function of the soil's water content, $K(w)$ (or of its matric potential, $K(\psi)$), is known as the *Richards equation*:

Because of the strong nonlinearity of the parameters involved (namely, the dependence of both the water content w and the hydraulic conductivity K on the soil's matric potential), the general flow equation for unsaturated conditions cannot be solved analytically, except in special cases. Numerical methods are used to represent what is fundamentally a continuous differential equation as an approximate algebraic equation (or set of equations) in which the domain of interest is cut into discrete intervals and derivates are replaced by difference equations that can be solved numerically by means of a computer for successive time increments.

Knowledge of the unsaturated hydraulic conductivity as a function of wetness and matric suction is generally required before any of the mathematical theories of water flow can be applied in practice. Methods have been proposed for predicting or approximating the necessary functions. Methods are also available for measuring the conductivity function directly, both in the laboratory and in the field. These methods are quite complex and exacting, and they are described in detail in more advanced textbooks and handbooks of soil physics.

SOLUTE TRANSPORT

Water present in the soil is never chemically pure. In the first place, the water entering the soil as rain or irrigation nearly always contains soluble constituents. Although rainwater is distilled and essentially pure when it first condenses to form clouds, as it then descends through the atmosphere, it generally dissolves such atmospheric gases as carbon dioxide and oxygen, often together with such products of our industrial civilization as oxides of sulfur and nitrogen, as well as—along the coasts—appreciable quantities of salt that enter the air as sea spray. Solute concentrations typically found in rainwater are of the order of 5 to 20 mg/L, units roughly equivalent to parts per million (ppm).

During its residence in the soil, the infiltrated water tends to dissolve additional solutes, including soluble products of mineral and organic matter decomposition as well as fertilizers and pesticides. The concentration of salts in irrigation water, generally obtained from surface reservoirs or subterranean aquifers, may range between 100 and 1000 mg/L. The concentration may be as high as 10,000 mg/L in drainage from saline soils. (Note: The concentration of salts in ocean water is of the order of 35 grams per liter.)

As it moves through the profile, soil water (more properly called "the soil solution") carries its solute load in its convective stream. The soil solution leaves some of it behind to the extent that the component salts are adsorbed by the soil's exchange complex, taken up by plants, or precipitated whenever their concentration exceeds their solubility, mainly at the soil surface during evaporation. Solutes move not only with the soil solution, but also within it, in response to concentration gradients. At the same time, solutes react among themselves and interact with the solid matrix of the soil in a continuous cyclic succession of interrelated physical and chemical processes. These interactions involve, and are strongly influenced by, such variable factors as acidity, temperature, oxidation-reduction potential, composition, and concentration of the soil solution.

The flux of solutes J_c carried by the convection (or *mass flow*, sometimes called the *Darcian flow*) of soil water is proportional to the flux of water q and to the concentration of the solutes c:

$$J_c = qc = -c[K(dH/dx)]$$

where $q = -K \, (dH/dx)$ is the Darcian flow discussed in a preceding section of this chapter. Since q is usually expressed as volume of liquid flowing through a unit area (perpendicular to the direction of flow) per unit time, and as c is the mass of solute per unit volume of solution, J_c is the mass of solute passing through a unit area of soil per unit time.

To estimate the distance of travel of a solute per unit time, we consider the average apparent velocity v of the flowing solution:

$$v = q/w$$

where w is the volumetric wetness and v is taken as the straight-line length of the path traversed in the soil in unit time. In this formulation, we ignore the roundabout path caused by the geometric tortuosity of the water-filled soil pores. Since actual velocities vary over several orders of magnitude within and between pores, the concept of average velocity is obviously a gross approximation. Accordingly, $J_c = v \, w \, c$.

The shortcoming of the above approach is that the transport of solutes seldom occurs by convection alone. Solutes do not merely move with the water as sedentary passengers in a train, but also move within the flowing water in response to concentration gradients, in the twin processes of *diffusion* and *hydrodynamic dispersion*. Moreover, solutes cannot generally be assumed to be inert, as they tend to interact with the biological and physicochemical systems within the soil.

Diffusion processes commonly occur within multicomponent fluids (gases and liquids), in consequence of the random thermal motion and repeated collisions and deflections of molecules and ions. The net effect is a tendency toward equalizing the spatial distribution of the components in the mixed fluid.

Diffusion processes are very important in the soil. In the air phase, the diffusion of such gases as oxygen, carbon dioxide, nitrogen, and water vapor strongly affect chemical and biological processes. Equally important are diffusion processes involving solutes in the soil's liquid phase. The processes include nutrient transfers to plant roots as well as the movement of various salts and of potentially toxic compounds that may affect soil biota.

Whenever solutes are not distributed uniformly throughout a solution, concentration gradients exist. Consequently, solutes tend to diffuse from zones where their concentration is higher to where it is lower. In bulk water at rest, the rate of diffusion J_d is related by *Fick's first law* to the gradient of the concentration c:

$$J_d = -D_0(dc/dx)$$

where D_0 is the diffusion coefficient for a particular solute diffusing in bulk water and dc/dx is the solute's effective concentration gradient.

The effective diffusion coefficient in the soil is generally less than in bulk water, since the liquid phase occupies only a fraction of the soil's volume, and the pores that it does occupy are tortuous so that the actual path length of diffusion is significantly greater than the apparent straight-line distance. In an unsaturated soil, as the fractional volume of water w diminishes, the tortuous path length actually increases. Accordingly,

$$D_s = D_0 w \xi$$

where ξ, the tortuosity factor, is an empirical parameter smaller than unity, expressing the ratio of the straight-line distance to the average roundabout path length through the water-filled pores for a diffusing substance. This parameter decreases with decreasing water content w.

The diffusion process is further complicated whenever there are sources or sinks for the diffusing substance within a given volume of soil. Such sources or sinks may result from the chemical or biological generation or absorption of diffusible substances along the path of diffusion.

Another process affecting solute movement in the soil is the phenomenon known as *hydrodynamic dispersion*. This process results from the microscopic nonuniformity of flow velocity in the soil's conducting pores during convection. Since water moves faster in wide pores than in narrow ones and faster in the center of each pore than along its walls, some portions of the flowing solution move ahead while other portions lag behind. The fact that some portions of the flowing solution move faster than other portions causes an incoming solution to mix or disperse within the antecedent solution. The degree of mixing depends on such factors as average flow velocity, pore-size distribution, degree of saturation, and concentration gradients. When the convective velocity is sufficiently high, the relative effect of hydrodynamic dispersion can greatly exceed that of molecular diffusion, and the latter can be neglected in the analysis of solute movement. On the other hand, when the solution is at rest, hydrodynamic dispersion does not come into play at all.

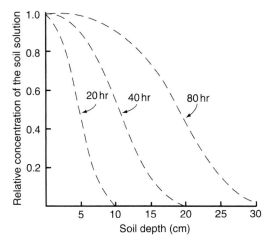

Fig. 7.5. Solute concentration vs. depth during infiltration of saline solution into non-saline soil. Note that the infiltrating "front" becomes increasingly diffuse about its mean position as it advances in the soil.

Mathematically, hydrodynamic dispersion can be formulated in a manner analogous to diffusion, except that a dispersion coefficient D_h is used instead of a diffusion coefficient. D_h has been found to depend more or less linearly on the average flow velocity. Because of the similarity in effect, though not in mechanism, between diffusion and dispersion, the two can sometimes be assumed to be additive. Accordingly, the diffusion and dispersion coefficients are sometimes combined into a single term, called the *diffusion-dispersion coefficient*, D_{sh}, which is a function of both the fractional water volume w and the average velocity v:

$$D_{sh}(w, \bar{v}) = D_s(\theta) + D_h(\bar{v})$$

A final comment regarding the flow of water and solutes in the soil: Although any given soil body (and the water within it, called the *soil solution*) may never attain complete equilibrium (chemical and/or physical) with its surroundings or even within itself (owing to repeated external or internal perturbations), its processes always trend spontaneously toward equilibrium. That is the principle expressed in the flow equations given above.

8. SOIL-WATER CYCLE

A land of hills and valleys, that drinketh
water as the rain of heaven cometh down
Deuteronomy 11:11

ENTRY OF WATER INTO SOIL

The soil plays an important role in the hydrological cycle. Particularly crucial to this role is the soil surface zone, where the interaction of atmospheric water takes place with the lithosphere. It is here that the complex partitioning of rainfall into infiltration, runoff, evaporation, transpiration, and groundwater recharge is initiated and sustained. This vital zone is also the primary site for the management and control of that all-important resource, water. The movement of water in the field can be characterized as a continuous, cyclic, repetitive sequence of processes, without beginning or end. For didactic purposes, we can describe this cycle as if it begins with the entry of water into the soil by the process of *infiltration*, continues with the temporary storage of water in the soil, and ends with its removal from the soil by drainage, evaporation, or plant uptake. Several fairly distinct stages of the cycle can be recognized.

When water is supplied from above to the soil, whether by rainfall or by irrigation, it typically penetrates the surface and is absorbed into successively deeper layers of the profile. The rate of infiltration, relative to the rate of water supply, determines how much water will enter the root zone and how much, if any, will accumulate over the surface or run off as *overland flow*. Hence, the rate of infiltration affects not only the water economy of terrestrial plants but also the amount of runoff and its attendant hazards of soil erosion and stream flooding.

The maximum rate at which the soil can absorb water that is applied to its surface at atmospheric pressure is called the soil's *infiltrability*. As long as the rate of water delivery, by rain or irrigation, is smaller than the soil's infiltrability, water infiltrates as fast as it arrives, and the process is *supply controlled*. However, when water is applied to the soil's surface at a rate exceeding the soil's infiltrability, the process becomes *soil-profile controlled*. The excess of supply rate over infiltrability will cause the extra water to either accumulate and pond over the soil surface if it is horizontal, or to run off the surface if it is sloping.

Measurements of infiltration rate into an initially dry soil under shallow ponding have shown that the soil's infiltrability is maximal at first and tends to decline over time, eventually approaching a constant rate that is the soil's *steady-state infiltrability*. The decrease of infiltrability from an initially high rate can in some cases result from gradual deterioration of soil structure and the partial sealing of the profile by the formation of a surface crust. It can also result from the detachment and migration of pore-blocking particles, from swelling of clay, and from entrapment of air bubbles or the bulk compression of the air originally present in the soil. Primarily, however, the decline of infiltrability results from the decrease in the matric suction gradient, (which occurs inevitably as infiltration proceeds and the soil in depth absorbs more water and, therefore, exercises less suction). The initial suction gradient is very strong between the infiltration-saturated surface zone and the relatively dry deeper layers. As the wetted zone deepens, the same difference in potential acting over a greater distance expresses itself as a diminishing suction gradient. As the wetted portion of the profile deepens, the suction gradient eventually becomes very small. In a horizontal column, the infiltration rate, initially high, tends to zero. In downward infiltration into a vertical column under continuous shallow ponding, the infiltration rate tends asymptotically to a steady, gravity-driven rate. That steady rate, in the case of a homogeneous and structurally stable soil, approximates the soil's hydraulic conductivity.

In the early stages of the process, the infiltration rate is strongly influenced by the profile's antecedent wetness, as an initially dry soil tends to draw water more strongly than an initially wet soil. As the process continues, the difference diminishes as the rate of infiltration approaches the steady rate characteristic of a deeply wetted profile.

The process of infiltration under rainfall is similar to infiltration under shallow ponding if the supply rate of water to the surface exceeds the soil's infiltrability. If rain intensity is less than the initial infiltrability value of the soil but greater than the final value, then at first the soil will absorb the water as fast as it arrives. However, if rain continues at the same intensity, and as soil infiltrability declines, the soil surface will eventually become saturated. Thenceforth the process will continue as in the case of ponding infiltration.

If the rain intensity at all times is lower than the soil's infiltrability (i.e., lower than its saturated hydraulic conductivity), the soil will continue to absorb the water as fast as it arrives without ever attaining saturation. Indeed, the lower the rain intensity, the lower the degree of saturation of the soil's wetted profile will be. This is indeed the principle applied in the modern technique of high-frequency low-intensity drip irrigation, which aims to maintain optimal soil aeration and moisture by avoiding soil saturation.

A rainstorm of any considerable duration typically consists of spurts of high-intensity rain punctuated by variable periods of low-intensity rain. During

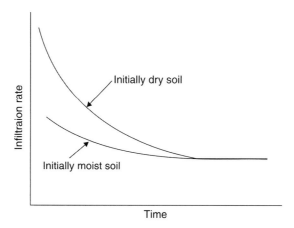

Fig. 8.1. Infiltrability as function of time in initially dry and initially moist profiles.

Table 8.1 Steady Infiltration Rates for Different Soil Types

Soil type	Steady infiltration (mm/hr)
Sand	>20
Sandy loam	10–20
Loam	5–10
Clay loam	1–5
Sodic clay	<1

such respite periods, surface soil moisture tends to diminish because of internal drainage, thus reestablishing a somewhat higher infiltrability. The next spurt of rain is therefore absorbed more readily at first, but soil infiltrability then continues to diminish as the soil is wetted to progressively greater depths and its suction gradients diminish. A complete analysis of the process under variable-intensity rain also entails the phenomenon of hysteresis, so it can be quite complicated.

When rainfall intensity exceeds the rate of infiltration, free water begins to accumulate over the soil surface if it is horizontal or pitted and to run off the surface if it is sloping. That so-called *overland flow* accelerates downslope, and as it does, it also becomes more erosive. Its capacity to detach soil particles and to carry them downslope increases with the water's velocity and turbulence.

Experimental studies have shown that the infiltration process can be greatly affected by soil-profile heterogeneity. The presence of a dense or fine-grained layer at any depth in the profile can retard the progress of infiltration after the wetting front reaches that layer, and can thenceforth cause the overlying layers to become saturated as, to borrow an analogy, in the case of a traffic jam backed up by a roadblock.

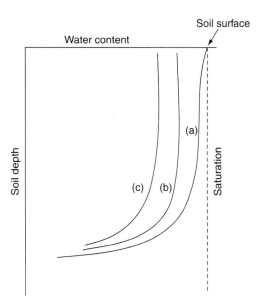

Fig. 8.2. The water content profile during infiltration: (a) under ponding; (b) under high-intensity sprinkling; and (c) under low-intensity sprinkling.

A very important special case of a layered soil is that of a profile that develops a thin, less permeable "crust" or "seal" at the surface.

The foregoing description of infiltration into various types of soil profiles has been based on the implicit assumption that the process occurs through the entire body of the soil, which absorbs and conducts water downwards at a variable, generally diminishing, rate. Where water permeates the porous network of the soil matrix and moves through its entire volume, we refer to the process as *distributed flow*. There are, however, anomalous cases in which the process takes place through distinct pathways that constitute only a fraction of the soil's total volume, while bypassing the remainder. Such cases can be characterized as *preferred pathway flow*, *bypass flow*, and in some cases *unstable flow*.

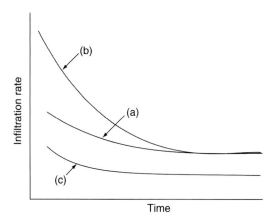

Fig. 8.3. Infiltrability as function of time in: (a) a uniform soil; (b) a soil with a more permeable upper layer; and (c) a soil with a surface crust.

BOX 8.1 On Rainfall

The important characteristics of a rainstorm are its time-variable intensity and its distribution of drop sizes, both of which affect the potential infiltrability and erosivity of the rain in relation to the soil. By *infiltrability*, we refer to the rain volume that can infiltrate the soil per unit area and time and, *ipso facto*, the fractional rain volume that will form surface water excess and tend to trickle off the surface. By *erosivity*, we refer to the cumulative energy of the raindrops striking the surface and likely to cause erosion. These parameters can refer to a single rainstorm or to the succession of rainstorms and their time distribution during the rainy season. A given season can be characterized in terms of its length, the number and frequency or timing of storm events, the intensity and duration of each event, etc. The important characteristics of the soil with respect to the rainfall regimen are its time-variable intake rate and its proneness to erosion, called its *erodibility*.

SURFACE RUNOFF

Whenever the rate of water supply to the soil surface exceeds the rate of infiltration, free water, called *surface water excess*, appears at the soil surface. Where the surface is not perfectly flat and smooth, the excess water collects in depressions, forming puddles. The total volume of water thus held per unit area is called the *surface storage capacity*. When the surface storage is filled and, if rainfall intensity continues to exceed the soil's infiltrability, the puddles begin to overflow and runoff may begin over sloping ground.

The term *surface runoff* thus represents the portion of the rainfall that is neither absorbed by the soil nor accumulates on its surface, but that runs downslope. Surface runoff typically begins as sheet flow, but as it accelerates and gains erosive power, it tends to scour the soil surface and to create channels, called *rills*. Running water gradually deepens and widens the rills, eventually forming a pattern of converging channels called *gullies* that finally discharge into streams and rivers.

REDISTRIBUTION AND RETENTION OF SOIL MOISTURE

When rain or irrigation ceases and free water on the surface disappears by absorption, runoff, or evaporation, the process of infiltration comes to an end. Downward water movement within the soil, however, does not cease immediately and may in fact persist for some time as soil moisture continues to percolate within the profile. The upper soil layers, wetted to near-saturation during infiltration, do not retain their full water content, as some of the water moves down into the lower layers under the influence of gravity and possibly also of suction gradients. If a water table exists at some level in the subsoil, the percolating water will tend to drain toward the water table in a process called *internal drainage*. If, however, a water table is absent, or if it is too deep to affect the water content of the soil profile, the post-infiltration movement of soil moisture is called *redistribution*. Its effect is to redistribute water in the soil by increasing the wetness of successively deeper layers at the expense of the infiltration-wetted upper layers of the profile.

In many cases, the rate of redistribution decreases rapidly as both the suction gradients and the hydraulic conductivity diminish and become practically imperceptible after several days. Thereafter, the initially wetted part of the profile appears to retain its moisture, unless it evaporates from the soil surface or is taken up by plants.

The importance of the redistribution process should be self-evident, because it determines the amount of water retained at various times in the different layers of the profile, and hence can affect the water economy of plants rooted in the soil. The rate and duration of downward flow during redistribution thus determine the effective *soil-water storage*. This property is vitally important, particularly in relatively dry regions, where water supply is infrequent and plants must rely for long periods on the unreplenished reservoir of water within the zone that can be tapped by roots. The redistribution process also determines how much water will flow through and beyond the root zone and how much leaching of solutes and recharge of groundwater will take place.

The fact that the post-infiltration downward movement of soil moisture slows down rapidly has led to the convenient approximation that the moisture in the initially wetted part of the profile actually ceases to drain after just two days and remains effectively constant thereafter. Accordingly, that moisture content is commonly termed the soil's *field capacity*. Though it is merely an approximation that is more realistic in some soils than in others, field capacity is a widely accepted measure of the soil's capacity to store moisture for subsequent plant use. The soils for which the field capacity concept is most applicable are coarse-textured (sandy) soils, in which the redistribution process is initially very rapid but soon slows down drastically owing to the steep decrease of hydraulic conductivity with increasing matric suction in the draining soil. In medium-textured or fine-textured soils, however, redistribution can persist at an appreciable rate for several more days.

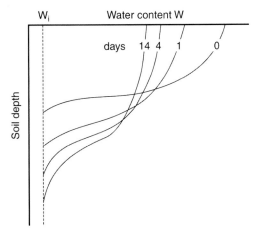

Fig. 8.4.   The changing moisture profile during redistribution. The successive moisture profiles shown are for 0, 1, 4, and 14 days after the cessation of infiltration. W_i shows the antecedent (pre-infiltration) soil wetness.

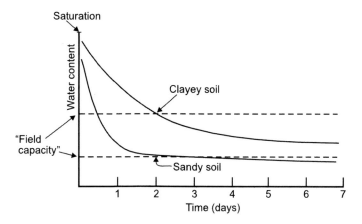

Fig. 8.5. The decrease of soil wetness with time in the initially wetted zone of a clayey and a sandy soil during redistribution. The dashed lines show the wetness values remaining in each of the soils after two days.

Table 8.2 Retention of Moisture (% by Mass) in a Medium-Textured Soil (0.6 to 0.9 m Depth) Following a Deep Wetting of the Profile

End of infiltration	29.0
After 1 day	20.2
2 days	18.7
7 days	17.5
30 days	15.9
60 days	14.7
156 days	13.6

THE WATER TABLE AND GROUNDWATER

Of the earth's total amount of fresh (nonsaline) water, some 75% exists in the frozen state in polar ice caps and glaciers, while less than 2% is in surface waters such as lakes and streams, and a relatively minute amount is contained in the generally unsaturated soil. That leaves nearly 22% of our planet's fresh water that permeates porous rocks and sediments, generally at some depth below the ground surface. It is this amount that we call *groundwater*. Soil and rock strata that contain groundwater and from which water can be extracted for human use are called *aquifers* ("water carriers").

In many regions, groundwater constitutes an important source of fresh water for domestic, agricultural, or industrial use. Since injudicious exploitation of groundwater can deplete the groundwater aquifer and diminish its quality, it is important to acquire knowledge regarding the behavior of groundwater and the best means of managing it. Extraction of groundwater in excess of the annual recharge— including natural percolation through the soil, as well as seepage from reservoirs and streams—will cause aquifer depletion, whereas

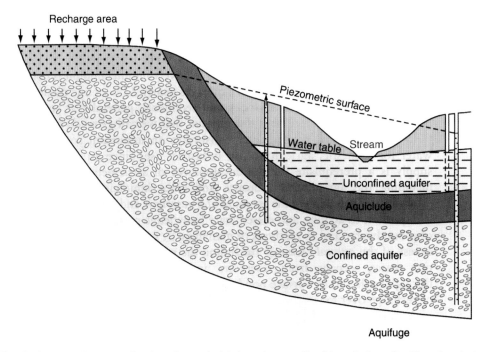

Fig. 8.6. A shallow unconfined aquifer, underlain by a deep confined (artesian) aquifer. Note that the two aquifers are separated by an impervious layer (aquiclude).

an excess of recharge over extraction will cause a buildup of groundwater and may even lead to waterlogging of soils.

Whereas water in an unsaturated soil is affected by suction gradients and its movement is subject to variations in conductivity resulting from changes in soil wetness, groundwater by definition is always under positive hydraulic pressure, and its pore spaces are saturated with water. Despite the differences between the saturated and unsaturated zones, the two are not independent realms but parts of a continuous flow system. Groundwater is recharged by percolation through the unsaturated zone, and the position of its surface, called the *water table*, is determined by the relative rates of recharge versus discharge. The water table rises wherever the recharge from the soil above exceeds the discharge below, and it descends wherever the opposite occurs. Just as water flow in the soil affects the position of the water table, the latter reciprocally affects the moisture profile and flow conditions above it.

Contrary to its name, the water "table" is hardly ever flat or level and may exhibit steep gradients. Where the topography of the land is variable, as well as where the inflow from precipitation or from stream seepage varies areally, the water table can change in depth and may in places and at times intersect the soil surface and emerge as free water (e.g., the out-seepage of a spring) or rise and permeate the entire soil profile (as in the case of marshes or wetlands) .

Soil saturation *per se* is not necessarily harmful to all plants. The roots of some specialized plants, especially those of water-loving plants, called *phreatophytes*, can even thrive in a saturated medium, provided it is free of toxic substances and contains or conducts sufficient oxygen to allow root respiration. However, most terrestrial plants are unable to transfer the required flux of

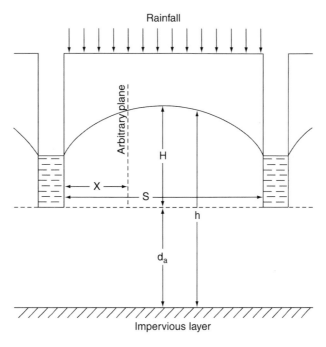

Fig. 8.7. Schematic cross section of the position of the water table in a ditch-drained field.

oxygen from their canopies to their roots internally. Excess water in the soil tends to block soil pores and thus to retard soil aeration; i.e., the free exchange of O_2 and CO_2 between the soil and the external atmosphere. The roots of many crop plants, generally classed as *mesophytes*, suffer greatly from restricted soil aeration and may fail to function unless the soil is drained of excess moisture.

In large areas, therefore, artificial drainage is required for the long-term maintenance of soil productivity. The common methods of drainage of water-logged soils is to dig ditches to below the water table, or to insert sectioned or perforated pipes below the water table, so as to lower it and to lead the excess water to some outlet. The water discharged from the drainage system must be disposed of in a way that does not harm the environment, especially if the drainage water is saline or contains harmful agents such as pesticide residues.

EVAPORATION FROM BARE SOIL

Incoming solar radiation and the sweeping action of wind typically impose an evaporative demand on the soil and on vegetation growing in it. If a field is bare of vegetation, evaporation takes place entirely from the soil surface. This process causes considerable losses of moisture from the upper layers of the soil, thus hampering the germination of seeds and the early growth of seedlings.

Three conditions are necessary for evaporation to occur and persist. First, there must be a continual supply of heat to meet the latent heat requirement for the conversion of liquid water into vapor. This heat can come from the body itself, thus causing it to cool, or—as is more common—it can come from the outside in the form of radiated energy from the sun or transmitted heat from the warm atmosphere. Second, the vapor pressure in the overlying atmosphere must remain lower than the vapor pressure at the surface of the evaporating body, and the vapor must be transported away by diffusion or convection. These two conditions—supply of energy and removal of vapor—are generally external to the evaporating body and are influenced by meteorological factors such as the solar radiation and the atmospheric temperature, humidity, and wind speed, which together determine the *atmospheric evaporativity*, the maximal rate at which the atmosphere can vaporize water from a free-water surface.

The third condition for evaporation to be sustained is that there be a continual supply of water from the interior of the soil body to the surface or the zone where evaporation takes place. This condition depends on the content and energy potential of water in the body as well as on its conductive properties; together, these properties determine the maximal rate at which water can be transmitted from the interior of the soil to the site of evaporation. Accordingly, the actual evaporation rate is determined by either the *evaporativity* of the atmosphere or by the *evaporability* of soil moisture (i.e., the soil's own ability to deliver water), whichever is the lesser (and hence the limiting factor).

If the top layer of soil is initially quite wet, as it typically is at the end of an infiltration episode such as a rainstorm, the process of evaporation will generally reduce soil wetness and thus increase the matric suction of the soil at the surface. This, in turn, will cause soil moisture to be drawn upward from the layers below, provided they are relatively wet and conductive. Among the conditions under which evaporation may take place are the following:

1. A shallow groundwater table may be present within the soil profile, or it may be absent (or too deep to affect evaporation at the surface). Where the water table occurs close to the surface, continual flow may take place by capillary rise from the saturated zone beneath through the unsaturated soil to the surface. If this flow is more or less steady, the process may persist, and the surface may remain moist for a considerable period of time. (Such is often the case in river valleys, where the continuous evaporation tends to deliver salts to the upper layers of the soil and cause it to become saline.) In the absence of a shallow water table, however, the loss of water at the surface will tend to dry out the soil gradually, from the top downwards.

2. The soil profile may be uniform in depth, or—more typically—it may consist of distinct layers differing in texture and properties; the profile may also be shallow or deep and with or without cracks that form secondary evaporation planes.

3. The soil surface may be bare or be covered by a layer of mulch (plant residues) differing from the soil in hydraulic, thermal, and diffusive properties.

4. Finally, external evaporativity may be constant, or it may fluctuate regularly (diurnally and seasonally) or irregularly. The evaporation process may be interrupted by episodes of rewetting by intermittent rainfall or by irrigation.

Where the evaporation continues for a considerable period and soil moisture is not replenished, the process results in a gradual drying of the soil. If external conditions, and hence atmospheric evaporativity, are more or less constant, the drying process typically occurs in three more-or-less distinct stages:

1. An initial *weather-controlled stage* that occurs while the upper layers of the soil are still wet and conductive enough to supply water to the surface at a rate commensurate with the evaporative demand (also called the *potential evaporation*) imposed by the sun and atmosphere. As such, this stage is largely independent of the properties of the soil profile, although it may be influenced by such properties of the soil surface as its reflectivity and the possible presence of a mulch over the surface that may act as a diffusion barrier. In an arid climate, this initial stage of evaporation is generally brief and may last only a few hours to a few days.

2. An intermediate *profile-controlled stage*, during which the evaporation rate falls progressively below the atmosphere's potential evaporativity. At this stage, the rate of the process is limited by the diminishing ability of the gradually drying soil profile to deliver moisture toward the evaporation zone. Hence, it is also called the *falling-rate stage*.

3. A residual, slow-rate, *vapor-diffusion stage* that may persist at a nearly steady rate for many days, weeks, or even months. This stage comes about after the surface zone has become so desiccated that it ceases to conduct liquid water but transmits only the slow diffusion of vapor. Some soils are characterized by a loose assemblage of aggregates or of loose sand that may dry quickly and become, in effect, "self-mulching".

Whereas the transition from the first to the second stage is generally a sharp one, the second stage generally blends into the third stage so gradually that the last two cannot be distinguished readily. As the soil surface gradually approaches equilibrium with the overlying atmosphere, it becomes more-or-less "air-dry". The zone of air-dryness becomes thicker in time, and the zone of evaporation becomes more diffuse, though the deepest layers of the profile may remain relatively moist for a very long time, even in a desert environment. In areas such as near seacoasts where there are appreciable day-night fluctuations of surface temperature and atmospheric humidity, the soil surface itself may also exhibit a pattern of fluctuating moistness due to night-time condensation of dew that is quickly re-evaporated during the following morning.

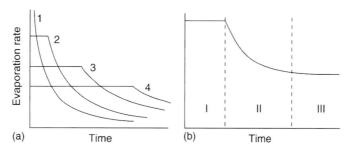

Fig. 8.8. Actual evaporation rate vs. time: (a) Under different potential evaporativity rates; (b) The three stages of the process.

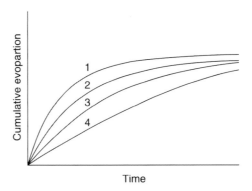

Fig. 8.9. Cumulative evaporation vs. time under different potential evaporativity rates.

PLANT UPTAKE OF SOIL MOISTURE

Green plants are *autotrophs*, able to create new living matter from inorganic raw materials. The major physiological function of growing plants is the process of *photosynthesis*, by which they combine water absorbed by the roots from the soil with carbon dioxide absorbed by the leaves from the atmosphere. That process, accompanied by the conversion of solar radiation into chemical energy, is summarized by the deceptively simple formula:

$$6CO_2 + 6H_2O + \text{sunlight energy} \rightarrow C_6H_{12}O_6 + 6O_2$$

We, along with the entire animal kingdom, owe our lives to this process, which not only produces our food but also releases into the atmosphere the elemental oxygen we need for our respiration. Plants also respire, and the process of respiration represents a reversal of photosynthesis, in the sense that some of the photosynthetic products, sugars and starches, are reoxidized and decomposed to yield energy and the original constituents:

$$C_6H_{12}O_6 + 6O_2 \rightarrow 6CO_2 + 6H_2O + \text{thermal energy}$$

In examining these formulas, we note immediately the central role of water as a major metabolic agent in the life of plants, as a source of hydrogen atoms for the reduction of carbon dioxide in photosynthesis, and as a product of respiration. Water is also the solvent and hence conveyor of ions and compounds into, within, and out of the plants. It is, in fact, a major structural component of plants, often constituting 90% or more of their total "fresh" mass. Much of this water occurs in cell vacuoles under positive pressure that keeps the cells turgid and gives rigidity to the plant as a whole.

Absorption of CO_2 from the atmosphere, where its concentration is less than 0.04%, takes place via small openings in the leaves called *stomates*. While those stomates are open, they are subject to the same evaporative demand of the atmosphere that tends to evaporate water from bare soils as described above. The evaporation of water from leaves is termed *transpiration*, and the sum of evaporation from the plants and the soil is termed *evapotranspiration*.

Only a small fraction of the water normally absorbed by plants growing in dry weather is used in photosynthesis, while most (often more than 95%) is lost as vapor in the process of transpiration. This process is impelled by the exposure to the atmosphere of a large area of moist cell surfaces, necessary to facilitate absorption of carbon dioxide and exchange of oxygen. In a sense, a plant growing in the field can be compared to a wick in an old-fashioned kerosene lamp. Such a wick, its bottom dipped into a reservoir of liquid (fuel) while its top is subject to the fire that burns and vaporizes it, must constantly transmit the liquid from bottom to top under the influence of physical forces imposed on it by the conditions prevailing at both its ends. To be sure, this is a very simplistic analogy, since plants are not passive and in fact are able at times to limit the rate of transpiration by shutting their stomates. However, for this limitation of transpiration most plants pay, sooner or later, in reduced growth, since the same stomates can no longer absorb carbon dioxide needed for photosynthesis. Reduced transpiration may also result in warming of the plants and hence in their increased *respiration* (the reverse of photosynthesis) and further reduction of *net photosynthesis*; i.e., the quantitative summation of assimilation minus respiration.

To grow successfully, a plant must achieve a water economy such that the demand made on it for water is balanced by the supply available to it. The problem is that the evaporative demand by the atmosphere is practically continuous, whereas the supply of water to the root zone by rainfall occurs only occasionally and irregularly. To survive during dry spells between rains, the plant must rely on the diminishing reserves of water contained in the pores of the soil, which itself loses water by direct evaporation and internal drainage.

The ability of a soil to store water and make it available to plants depends on soil texture and profile characteristics. Sandy soils allow infiltrated water to drain to great depth while retaining only a small amount within the root zone. On the opposite end of the textural spectrum, clayey soils may retain rain or irrigation water near the surface, where the excessively wet layer may restrict aeration at first and then lose much of the water to evaporation, especially as clay tends to form cracks. An optimal soil is of intermediate loamy texture that absorbs water throughout the root zone without too much loss due to deep percolation. Ideally, such a soil should have an aggregated, porous top-layer that absorbs rain readily but that, as it dries, forms a barrier to excessive evaporation during dry spells between rains. The presence of a mulch of plant residues over the surface can help to promote infiltration as well as to restrict evaporation.

In many areas, in fact, the total water requirements of plants during their main growing season tend to exceed the seasonal rainfall. The relation

BOX 8.2 Structural Adaptation of Plants

Terrestrial plants are structured to maximize their surface exposure, both above and below the ground surface. Their aerial canopies frequently exceed the area of covered ground many times over. Such a large surface helps the plants to intercept and collect sunlight and carbon dioxide, two resources that are normally diffuse rather than concentrated. Even more striking is the shape of roots, which proliferate and ramify throughout a large volume of soil while exposing an enormous surface area. A single plant can develop a root system with a total length of several kilometers and with a total surface area of several hundred square meters. The need for such exposure becomes apparent if we consider the primary function of roots, which is to gather water and nutrients continuously from a medium that often holds only a meager supply of water per unit volume and that contains soluble nutrients only in very dilute concentrations. And while the atmosphere is a well-stirred and thoroughly mixed fluid, the soil solution is a sluggish and unstirred fluid that moves toward the roots at a slow pace, so the roots must move toward the moisture by constantly growing and foraging through as large a volume of soil as possible. The growth of roots in the soil is affected by a host of factors additional to moisture and nutrients, including temperature, aeration, mechanical resistance, acidity or alkalinity, salinity, and the possible presence of various toxic substances.

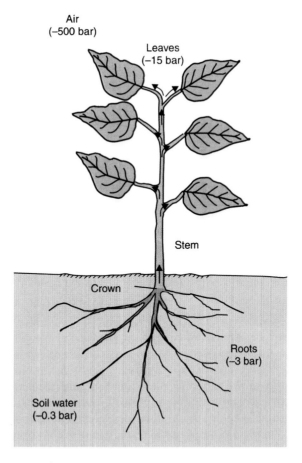

Fig. 8.10.   Variation of water potential along the transpiration stream.

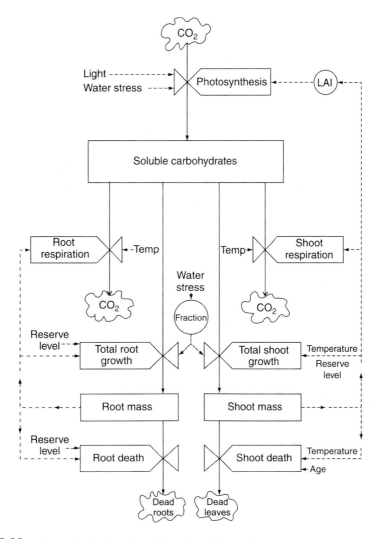

Fig. 8.11. The cycle of carbon in a living plant (schematic).

between demand and supply varies from region to region. In a moist region, the ratio between the amount of water transpired by plants and the amount of dry matter they produce (a ratio often taken as an index of "water-use efficiency") may be of the order of 100 to 300. That ratio may be as high as 1000 or so in an arid region, where the dry and windy air and intense solar radiation combine to increase the demand for water even as the paucity of rainfall limits the supply. Hence, there is a need to augment that supply by artificial irrigation in arid regions wherever water can be made available from streams or aquifers.

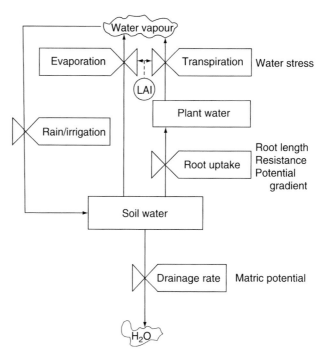

Fig. 8.12.   The cycle of water in the soil-plant-atmosphere continuum (schematic). Note: LAI is the leaf-area index of the vegetation.

SOIL-PLANT-ATMOSPHERE SYSTEM AS A BIOPHYSICAL CONTINUUM

Current approaches to the topic of soil-water extraction and utilization by plants are based on recognition that the field with all its components—soil, plant, and ambient atmosphere taken together—constitutes a physically integrated, dynamic system in which all flow processes occur interdependently like links in a chain. This unified system has been called the *soil-plant-atmosphere contin-uum* (SPAC for short). The universal principle operating consistently throughout the system is that water flows spontaneously from regions where its potential

---

BOX 8.3   Water Relations of Plants

The stomates that are open to absorb carbon dioxide from the atmosphere in the process of photosynthesis and exchange oxygen in the process of respiration also release water vapor from the plant to the atmosphere in the process of transpi-ration. During the active growth phase of many mesophytes (which include most crop plants), water may constitute some 90% or more of the plant mass. However, the daily transmission of water to the atmosphere in the process of transpiration may exceed the plants' water content several-fold. In a typical growing season, many crop plants retain less than 1% of the water they absorb from the soil. The rate of transpiration, however, is highly variable, as it depends on the prevailing climate.

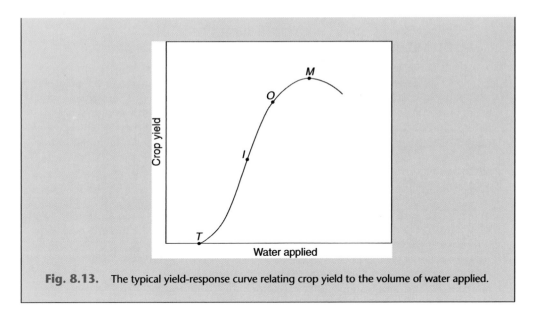

Fig. 8.13. The typical yield-response curve relating crop yield to the volume of water applied.

energy state is higher to where it is lower. Indeed, it is the gradient of that energy potential that constitutes the force inducing flows within and between the soil, the plant, and the atmosphere. This principle applies even though the various components of the overall potential gradient are effective to varying degrees in motivating flow in different sections of the soil-plant-atmosphere system.

The flow path in the SPAC includes liquid water movement in the soil toward the roots, liquid, and perhaps vapor movement across the root-to-soil contact zone (in some cases mediated by fungal growths), absorption into the roots and across their membranes to the vascular tubes of the xylem, transfer through the xylem up the stem and branches to the leaves, evaporation in the intercellular spaces within the leaves, vapor diffusion through the sub-stomatal cavities and out the stomatal apertures to the boundary air layer in direct contact with the leaf surfaces, and through it, finally, to the turbulent atmosphere, which sweeps away most of the water extracted by the plant from the soil.

Also included in that complex series of catenary processes are interactions involving various solutes. Some of the solutes present in the soil solution are excluded by the plant roots (and hence tend to concentrate in the peripheries of water-extracting roots, from where they tend to diffuse outwards), and some, particularly nutrients, are absorbed into the roots and thence transferred to all plant parts. Other interactive processes are such that involve respiration and gaseous exchanges between plants, as well as other organisms in the soil community, and their environment, greatly influenced by energy flows in radiant and thermal (latent and sensible) forms. Imagine all that, and more, taking place at variable rates and directions throughout the soil-plant-atmosphere system continuously during the diurnal and seasonal cycles, and marvel at the intricacy and dynamism of life in the soil!

BOX 8.4   Wetlands

Wetlands are tracts of land that are saturated with water more or less permanently. Wetlands usually occur in low-lying areas, alongside the banks or estuaries of rivers, or the shores of lakes or seas. Such marshy areas serve as environmental filters, where dissolved or suspended matter (including nutrients such as nitrogen and phosphorus, particulate sediments, and various pollutants) deposited by incoming waters are retained, and relatively pure water is released. The maintenance of flood-absorbing wetlands in river valleys can help protect downstream areas from periodic flooding, which can otherwise be very damaging.

Where wetlands are drained, the pollutants they would normally tend to filter may flow directly into lakes and rivers, causing their excessive enrichment with nutrients. As a result of such enrichment, called *eutrophication*, aquatic plants and algae tend to proliferate and to deplete the water bodies of oxygen, thus killing the fish and possibly turning clear-water lakes into turbid, foul-smelling morasses.

A possibly negative aspect of wetlands is their tendency to emit methane (natural gas), a product of the anaerobic decomposition of organic matter in the saturated soil. Methane is a radiatively active gas that contributes to the atmospheric "greenhouse effect" and thus to the process of global warming. On the other hand, the drainage of wetlands generally results in the rapid aerobic decomposition of accumulated organic matter and thus to the release of carbon dioxide, which itself is a contributor to global warming, albeit a less powerful one than methane.

BOX 8.5   Soil as a Medium for Seed Germination and Root Growth

For optimal germination and seedling establishment, seeds must be embedded in firm contact with a soil that is moist but also a well-aerated, be placed deep enough to avoid desiccation but not too deep to emerge, be in a medium soft enough to avoid mechanical impedance, be at a favorable temperature, and be without excess salts, acidity, or other toxic factors. To allow continued plant growth and maturation, the soil must also be deep and porous enough to allow firm but permeable anchorage to proliferating roots, and provide a steady availability of water and nutrients along with a continual supply of oxygen for respiration and disposal of carbon dioxide.

As the root system supplies the plant with water and nutrients, the important variables that affect its function are its depth and lateral extent (i.e., the volume of soil tapped) and the density of the roots per unit volume of soil. Although they grow and proliferate continuously during the plant's active growing stage, even at maximal density the roots occupy only a small fraction (typically between 1 and 5%) of the soil volume in the rooting zone. Hence, the plants also depend on the conduction of water and solutes in the soil surrounding the roots.

The typical growth curve of a young plant proceeds at first at an exponentially rising rate that tends to become nearly constant as new leaves shade over older leaves, then begins to diminish as the plant approaches maturity. The growth rate falls to zero as the plant reaches its limited size, and it becomes negative as the plant senesces and eventually dies.

9. SOIL-WATER AND SOIL-ENERGY BALANCES IN THE FIELD

The knowledge of man is as the waters,
some descending from above,
and some springing from beneath.
Francis Bacon (1561–1626)

WATER AND ENERGY RELATIONS IN THE SOIL

The various soil-water flow processes described in the preceding chapters as separate phenomena—infiltration, redistribution, drainage, evaporation, and uptake of water by plants—are in fact strongly interdependent, as they occur sequentially or simultaneously. In this chapter, we shall endeavor to integrate those processes by summation of their mass and energy balances.

The field water balance, like a financial statement of income and expenditures, is an account of all quantities of water added to, subtracted from, and stored within a given volume of the soil (e.g., the rooting zone) during a specified period of time. As such, the water balance is based on the law of conservation of mass, which states that matter is neither created nor destroyed but is only changed from one state or location to another. Assuming that no significant amounts of water are either composed or decomposed in the soil, the water content of a soil body of finite volume cannot increase appreciably without addition from the outside—as by infiltration from above, capillary rise from below, or condensation from incoming air—nor can it be diminished unless transported to the atmosphere by evapotranspiration or to deeper zones by drainage.

The field water balance is intimately associated with the energy balance, since it involves processes that require, and are driven by, energy. The energy balance is an expression of the law of conservation of energy, which states that, in a given system, energy can be absorbed from, or released to, the surroundings and that it can change form, but it cannot be created or destroyed.

The content of water in the soil affects the way the energy flux reaching the field is partitioned and utilized. Likewise, the energy flux with its transformations affects the state and movement of water. So the two balances are inextricably interlinked. In particular, the evapotranspiration process, which depends on the simultaneous supply of both water and energy, is often the largest component of the water balance, as well as the largest component of the energy balance in the field.

Water Balance of the Root Zone

The volume of soil for which the water balance may be calculated is determined arbitrarily, depending on the purpose for which the data is to be used: it may encompass a small field plot or an entire watershed.

In its simplest form, the water balance states that any change in the water content (ΔW) of a given body of soil during a given period must equal the difference between the amount of water added to, and the amount withdrawn from, that body:

$$\Delta W = W_{in} - W_{out}$$

When the gains exceed the losses, the water content change is positive; conversely, when the losses exceed the gains, ΔW is negative. The soil volume of interest is thus regarded as a bank account or a storage reservoir.

To itemize the accretions and depletions from the soil storage reservoir, one must consider the disposition of rain or irrigation reaching a unit area of soil surface during a period of time. Some of the rain infiltrates the soil, and some

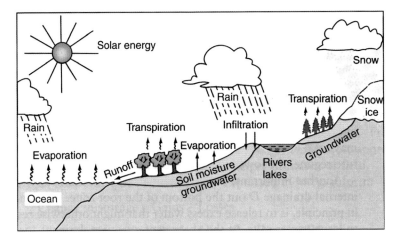

Fig. 9.1. The hydrologic cycle.

may exit the area as overland flow (surface runoff). Of the water infiltrated, some may evaporate directly from the soil surface, some may be taken up by plants for growth or transpiration, some may drain downward beyond the root zone, and the remainder accumulates within the root zone and adds to the soil moisture content. Additional water may reach the given soil body by runoff from a higher area or by capillary rise from below.

The root-zone water balance is usually expressed in integral form:

$$\text{Change in storage} = \text{Gains} - \text{Losses}$$
$$(\Delta S + \Delta V) = (P + I + U) - (R + D + E + T)$$

where ΔS is the change in soil moisture storage, ΔV is the increment of water in vegetative biomass, P is precipitation (rain and snow), I is irrigation, U is upward capillary flow into the root zone, R is runoff, D is downward drainage out of the root zone, E is direct evaporation from the soil, and T is transpiration by plants. All quantities are expressed in terms of volume of water per unit land area during the period considered.

The largest component in the "losses" part of the equation above is generally the evapotranspiration $E + T$. It is appropriate at this point to introduce the concept of *potential evapotranspiration* (designated PET), representing the climatic "demand" for water. PET from a well-watered field depends primarily on the energy supplied to the surface by solar radiation and on the transport of vapor away from the soil surface to the atmosphere by diffusion and advection. Hence, it is convenient to assume that PET depends primarily on the external climatic inputs and that it is relatively (though not absolutely) independent of the properties of the soil itself.

Actual evapotranspiration (AET) is generally a fraction of PET, depending on the degree and density of plant coverage of the surface, as well as on soil moisture and root distribution. AET from a well-watered stand of a close-growing crop will generally approach PET during the active growing stage, but the AET may fall below the PET during the early-growth stage, prior to full canopy coverage, and again toward the end of the growing season as the matured plants begin to dry out. For an entire growing season, AET may total 60–80% of PET, depending on the water supply: the drier the soil moisture regime, the lower the actual relative to the potential evapotranspiration.

An important item of the field water balance is the internal drainage D out of the bottom of the root zone. In humid areas, drainage releases excess water that might otherwise restrict soil aeration, especially in fine-textured soils. It also leaches away excess salts. Where drainage is inhibited, waterlogging, as well as salt accumulation—a particular hazard in arid zones—might occur. However, excessive drainage, such as may occur from sandy soils, may deprive the root zone of vital water and cause early wilting of sensitive plants.

Radiation Exchange in the Field

The term *radiation* refers to the emission of energy in the form of photons or electromagnetic waves from all bodies above 0 K. *Solar radiation* received on the earth's surface is the major component of its energy balance. Green plants are able to convert a part of the received radiation into chemical energy in the process of photosynthesis, on which all life on earth ultimately depends.

Solar radiation arrives at the outer surface of the atmosphere at a practically constant flux of nearly 1.4 joules per second per square meter perpendicular to the incident radiation (i.e., 1.353 kW/m², about 2 cal/min cm²). About half of this radiation is in the wavelength range of visible light (0.4–0.7 μm). The solar radiation spectrum corresponds approximately to the emission spectrum of a so-called "black body" at a temperature of 6000 K.

The Earth, too, emits radiation, but since its surface temperature is about 300 K, this *terrestrial radiation* is of much lower intensity and longer wavelength (3–50 μm) than solar radiation. Between the two spectra, the sun's and the earth's, there is very little overlap, so it is customary to refer to the first as *short-wave radiation* and to the second as *long-wave radiation.*

In passage through the atmosphere, solar radiation changes both its flux and its spectral composition. About one third of it, on average, is reflected back to space. (This reflection can be as high as 80% when the sky is completely overcast with clouds.) In addition, the atmosphere absorbs and scatters a part of the radiation, so only about half of the original flux density of solar radiation finally reaches the ground directly. A part of the reflected and scattered radiation also reaches the ground indirectly and is called *sky radiation.*

A fraction of the short-wave radiation reaching the surface is reflected by it. That fraction is known as the *albedo* (suggesting "whiteness") and is also referred to as the *reflectivity coefficient.* It depends on the color, roughness, and inclination of the surface and is of the order of 5–10% for water, 10–30% for vegetation, 15–45% for a bare soil, and up to 90% for fresh snow.

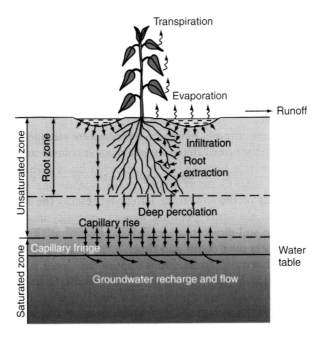

Fig. 9.2.   The water balance of a root zone.

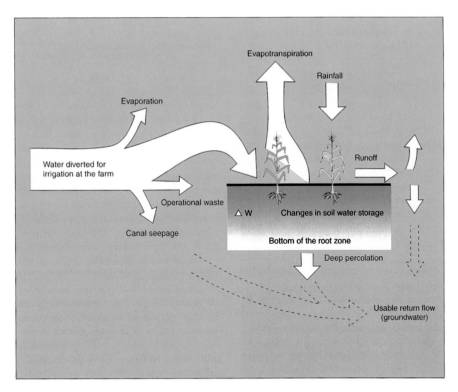

Fig. 9.3. The water balance of a field.

In addition to these fluxes of incoming and reflected short-wave radiation, there occurs an exchange of long-wave (heat) radiation. The earth's surface, which converts the short-wave radiation it absorbs to sensible heat, emits long-wave (infrared) radiation. At the same time the atmosphere also absorbs and emits long-wave radiation, part of which reaches the earth's surface. The difference between these outgoing and incoming fluxes is called the *net long-wave radiation flux*. During the day, the net long-wave radiation may be a small fraction of the total radiation balance, but during the night, in the absence of direct solar radiation, the heat exchange between the land surface and the atmosphere dominates the radiation balance.

The overall difference between total incoming and total outgoing radiation (including both the short-wave and the long-wave components) is termed *net radiation*, and it expresses the rate of radiant energy absorption by the field:

$$J_n = (J_s \downarrow - J_s \uparrow) + (J_l \downarrow - J_l \uparrow)$$

where J_n is the net radiation, $J_s \downarrow$ is the incoming flux of short-wave radiation from sun and sky, $J_s \uparrow$ is the short-wave radiation reflected by the surface, $J_l \downarrow$ is the flux of long-wave radiation incoming from the sky, and $J_l \uparrow$ is the long-wave radiation emitted by the surface. At night, the short-wave fluxes are negligible, and since the long-wave radiation emitted by the surface generally

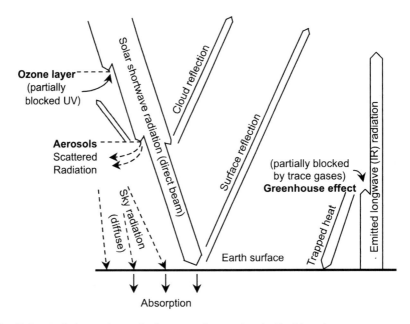

Fig. 9.4. Radiation exchange in the atmosphere and at the Earth's surface.

exceeds that received from the sky, the nighttime net radiation flux is negative, as the surface loses heat and may become colder than the overlying atmosphere.

The flux of short-wave radiation is equal to the product of the incoming short-wave flux and the reflectivity coefficient (the albedo α):

$$J_s \uparrow = \alpha J_s \downarrow$$

Therefore,

$$J_n = J_s \downarrow (1-\alpha) - J_1$$

where J_1 is the net flux of long-wave radiation and is given a negative sign. (Since the surface of the earth is usually warmer than the atmosphere, there is generally a net loss of thermal radiation from the surface.) As a rough average, J_n is typically of the order of 55–70% of $J_s \downarrow$.

Table 9.1　Reflectivity (Albedo) Values of Different Surfaces (% of Incoming Solar Radiation Flux)

Fresh snow	~90
Desert sand	35
Bare soil	10–30
Grassy meadow	20
Deciduous forest	15
Coniferous forest	10

Table 9.2 Thermal Properties of Soil Components (Approximate)

	Volumetric heat capacity (kJ/m3deg.)	Thermal conductivity (W/m.deg.)
Quartz sand	2000	8.8
Clay minerals	2000	3.0
Organic matter	2700	0.25
Water	4200	0.6
Air	~1.2	0.025

Total Energy Balance

Given the net radiation as the difference between the incoming and outgoing fluxes of radiant energy, we now consider the transformations of this energy in the field. Part of the received net radiation is transformed into heat, which warms the soil and the plants growing in it, and, in turn, warms the overlying atmosphere. Another part is taken up by the plants in their metabolic processes; e.g., photosynthesis. Finally, a major part is generally absorbed at latent heat in the twin processes of evaporation and transpiration. Thus,

$$J_n = LE + A + S + M$$

where LE is the rate of energy utilized in evapotranspiration (a product of the rate of water evaporation E and the latent heat of vaporization L); A is the energy flux that goes into heating the air (called *sensible heat*); S is the rate at which heat is stored in the soil, water, and vegetation; and M represents the miscellaneous other energy terms, such as are involved in the metabolic processes of photosynthesis and respiration.

Where the vegetation is short (e.g., grass or a field crop), the storage of heat in the vegetative biomass is negligible compared with storage in the soil. (The situation might be different in the case of a dense forest with massive trees.) The heat stored in the soil during the day, however, may be lost in large part during the night, when the flux reverses direction. For this reason, soil temperatures do not vary greatly from day to day, but may well vary appreciably from winter to summer. The process of photosynthesis rarely accounts for as much as 5% of the daily net radiation, and in most cases is much less than that. One constraint to the more efficient utilization of sunlight in photosynthesis is the generally low concentration of carbon dioxide in the air. Growing plants in an atmosphere artificially enriched with CO_2 may improve the utilization of sunlight, but such enrichment is generally practical only in enclosed greenhouses. The concentration of CO_2 in the open atmosphere has increased since the beginning of the Industrial Revolution by some 30% (from about 270 to about 380 parts per million), but the potential benefit of that increase may well be negated by the associated process of global warming, which tends to increase the rates of both respiration and transpiration.

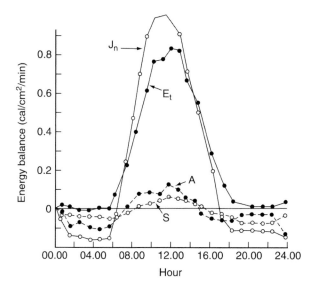

Fig. 9.5.   The diurnal variation of net radiation and of its utilization by evapotranspiration and heating of the atmosphere and soil.

Overall, the increments of energy stored in soil and vegetation and fixed photochemically account for a small portion of the total daily net radiation, with the greater portion going into latent and sensible heat. The relative allocation between these two terms depends on the availability of water for evaporation, but in most cases the latent heat predominates over the sensible heat term.

THE ATMOSPHERIC GREENHOUSE EFFECT

The earth's surface naturally tends to maintain an energy balance such that the total amount of radiant energy received is equal to the total amount released. Since the flux of emitted infrared radiation depends on the surface temperature, that temperature adjusts itself over time so that the balance is maintained. The radiation balance is a self-adjusting system: If more energy comes in than goes out, the surface becomes warmer so that more energy is then emitted, and vice versa.

According to the well-known *Stefan-Boltzmann law*, the energy emission rate from the surface of a body, integrated over all wavelengths, is proportional to the fourth power of the absolute temperature. If the surface temperature rises from, say, 10 C (283 K) to 30 C (303 K), it will increase its rate of heat release by $(303/283)^4$. or 31%.

If the Earth were without a gaseous atmosphere, calculations show that its mean surface temperature at equilibrium would be some 33 C lower than it actually is. The presence of the atmosphere causes the surface to be warm enough to promote the profusion of life as we know it. That effect of the atmosphere is called the *natural greenhouse effect*. The name is based on the

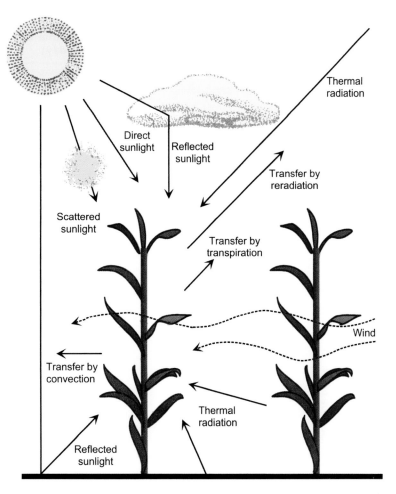

Fig. 9.6. Energy exchange between a plant stand and its environment (After Gates 1980).

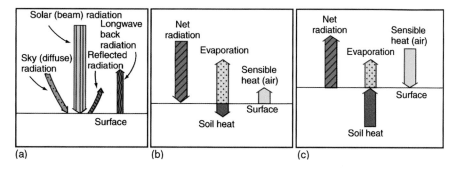

Fig. 9.7. Schematic representation of the radiation balance (a) and of the daytime (b) and nighttime (c) energy balance of a field.

Table 9.3   Daily Energy Balance of a Vegetated Field (Hypothetical Fluxes on a Clear Summer Day; Units: Megajoules per square meter per day)

Radiation:	
Incoming solar	30.0
Reflected solar	6.0
Outgoing emitted	10.0
Net	14.0
Energy uptake:	
Evapotranspiration	12.0
Sensible heat (air & plants)	1.0
Soil heat	0.8
Plant growth	0.2
Total	14.0

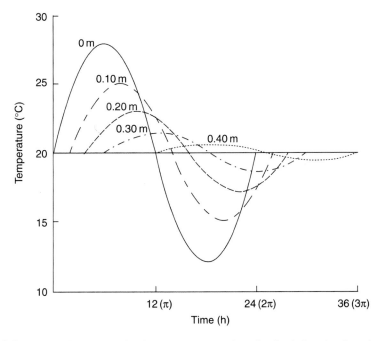

Fig. 9.8.   Idealized variation of soil temperature at various depths during the diurnal cycle.

perception that the effect of the atmosphere is analogous to the effect of a glass-covered greenhouse.

The interior of a greenhouse is typically warmer than the external environment, primarily because of the optical properties of its glass walls. Clear glass is nearly transparent to light; hence, it admits the incoming solar radiation practically without hindrance. Inside the greenhouse, that energy is converted to heat. However, the same glass that admits more than 90% of the incoming visible light absorbs some 90% of the infrared radiation at wavelengths greater than 2000 nm. So the release of heat by infrared radiation from the interior of the greenhouse is partially blocked by the glass. As the interior gets hotter, its flux of emitted radiation intensifies and its wavelength shortens, until the fraction transmitted by the glass compensates for the incoming radiation. A new energy balance is then established, at a higher surface temperature.

The atmosphere is not exactly like the walls of a greenhouse. It is thick, fluid, and turbulent rather than thin and rigid, so it mixes and exchanges heat and vapor with the environment in a way that glass cannot. But the greenhouse analogy is apt in one important respect: The atmosphere is preferentially permeable to short-wave solar radiation. Except for the suspended dust particles, which absorb solar radiation, and clouds, which reflect it, the clear atmosphere is largely transparent to incoming light. Not so to outgoing heat—although the major gases (nitrogen and oxygen, plus argon) that make up well over 99% of the atmosphere transmit infrared radiation readily, some of the gases present in small concentrations tend to absorb emitted infrared radiation. These radiatively active gases are commonly referred to as "greenhouse gases".

The most important greenhouse gas, being relatively abundant and ubiquitous—though in variable concentrations—is water vapor. The second most important is carbon dioxide, which is continually being cycled throughout the biosphere via such processes as photosynthesis, respiration, and decomposition of organic matter. The natural greenhouse effect, as far as it goes, is biologically beneficial because it creates conditions that promote life in regions that would otherwise be too cold for many organisms.

Ever since the beginning of the Industrial Revolution, little over two centuries ago, human activity has resulted in a significant infusion of carbon dioxide to the atmosphere, mainly by the burning of fossil fuels, coal and petroleum. Additional releases of carbon dioxide have resulted from the clearing of forests as well as from the cultivation of virgin lands, which hastens the decomposition of soil organic matter without commensurate replenishment. All these anthropogenic processes have already raised the concentration of CO_2 from its original level of about 270 parts per million by volume to some 370 ppmv.

Other greenhouse gases whose concentrations are also increasing due to human activity are: methane (CH_4), which is emitted from marshes, rice paddies, and the fermentative digestion of cattle; and nitrous oxide (N_2O), which results from the burning of fossil fuels and of biomass, as well as from fertilizer use. In addition to these natural gases, modern industry has also emitted synthetic gases known as *chlorofluorocarbons* (CFCs), which have been used as refrigerants, foaming agents, and propellants for spray cans. The manufacture and use of the latter gases, which tend to destroy the stratosphere's protective layer of ozone, has been curtailed by international agreement (the so-called Montreal Protocol of 1987), but not yet stopped completely.

Altogether, the human-enhanced greenhouse effect poses the danger of elevating the average temperature of the earth's surface by several degrees—perhaps by as much as 4 or 5 C by the middle of this century. Although CO_2 enrichment may spur photosynthesis and to some degree improve water-use efficiency by some crops, the enhanced greenhouse effect, on balance, may do more harm than good. The result might be not only a warmer earth and more extreme climate, but also a greater evaporational demand, a drier soil, and more extreme episodes of drought as well as more violent storms.

Soils and the Atmospheric Greenhouse Effect

The principal cause of the enhanced greenhouse effect is now known to be the burning of fossil fuels, coal and petroleum. An important additional source of greenhouse-gas emissions are terrestrial ecosystems, particularly the soil, as managed or mismanaged by humans. The burning of forests results in the immediate release of CO_2. The conversion of virgin soils to agricultural cultivation typically results in the gradual oxidation of soil organic matter and the emission of CO_2. Where soil aeration is impaired, methane (CH_4) and nitrous oxide (N_2O) are released, both being greenhouse gases in addition to CO_2.

The amount of carbon stored in soil organic matter has been estimated to total some 1.5×10^{18} grams (1500 billion tons). (In addition, soils contain carbon in the minerals compounds $CaCO_3$ and $MgCO_3$.)

Calculations suggest that prior to 1960 the release of CO_2 into the atmosphere from soils and vegetation had exceeded the release from the burning of fossil fuels. At present, the contributions to the atmosphere of carbon in the form of CO_2 from deforestation and land use are estimated to total 1.5 billion tons per year, which constitute nearly 25% of the total anthropogenic emissions of about 7 billion tons per year. The net emissions from soils alone are likely to be in the order of 0.5 billion tons per year.

The concentration of methane (CH_4) in the atmosphere prior to the Industrial Revolution, as measured in air bubbles trapped in glaciers, was of the order of 0.8 parts per million on the volume basis. The concentration in the year 2000 was approximately 1.8 ppmv. The emission of CH_4 (popularly known as "marsh gas") takes place mainly from anaerobic soils in wetlands (marshlands) and paddy fields. Additional releases of this gas are due to the digestive processes of domestic ruminants (cattle and sheep), as well as to leakages from the production and distribution of "natural gas".

Nitrous oxide (N_2O) is produced in anaerobic soils by the biological reduction of nitrate, a process called *denitrification*. The concentration of nitrous oxide (N_2O) in the atmosphere is about 320 ppbv and has been increasing

Table 9.4   Main Greenhouse Gases Influenced by Human Activity

	CO2	CH4	N2O	CFC-11	CFC-12
Pre-industrial (1750) atmospheric conc.	280 ppmv	0.8 ppmv	288 ppbv	0	0
Recent (1990) atmospheric conc.	353 ppmv	1.72 ppmv	310 ppbv	280 pptv	484 pptv
Annual rate of accumulation	1.8 ppmv (0.5%)	0.015 ppmv (0.9%)	0.8 ppbv (0.25%)	9.5 pptv (4%)	17 pptv (4%)
Atmospheric lifetime (years)	50–200	10	150	65	130

at the rate of about 0.25% per year. At this low concentration, it has only a marginal effect on global warming, but it has a relatively long residence time, about 150 years. As it rises to the stratosphere, it is oxidized to nitric oxide (NO), which tends to destroy ozone molecules.

A fuller description of the role of soils in the enhanced greenhouse effect is given in the Appendix.

10. SOIL CHEMICAL ATTRIBUTES AND PROCESSES

To the wise man, the whole world's a soil.
W. T. Kelvin, 1824–1907

SOIL ORGANIC MATTER

Terrestrial plants and animals normally deposit their residues in and on the soil, so that, in the natural state, the upper horizons of the soil (the so-called topsoil) generally accumulate significant amounts of organic matter. The total content of organic matter in the A horizon (say, the top 10 to 30 cm) of a virgin soil may vary from less than 1% of the total mass in the case of sparsely vegetated soils in an arid zone to 10% or more in the case of a densely vegetated soil in a humid zone. Waterlogged soils of marshes, where the vegetation is especially profuse and the rate of decomposition of organic residues is inhibited by lack of aeration, may contain over 50% organic matter. Altogether, the amount of organic carbon stored in the soils of the world and in the biota they support greatly exceeds the amount of carbon in the atmosphere, hence any changes in soil organic matter content can have a significant effect on the so-called "atmospheric greenhouse effect" and on global warming.

Organic matter in the soil is an exceedingly complex mixture of compounds that are responsible for much of the soil's bio-physico-chemical activity. Organic matter in the soil fulfills many functions. It serves as a growth medium and energy source for microorganisms that, in the process of decomposing organic matter, release nutrients for plants. Organic matter imparts stability to soil aggregates

135

and thus improves soil structure, infiltration, and aeration. Modern agriculture, which relies on mechanized power rather than on draft animals, too often results in depletion of organic matter. Hence, special measures are needed to augment the organic content of soils by means of composting and green manuring. The latter practice consists of the periodic planting of a legume (e.g., clover, vetch, cowpeas, lupines, etc.) not to be harvested but to be mixed into the topsoil for the purpose of augmenting topsoil carbon and nitrogen reserves.

Mature natural ecosystems generally attain an equilibrium content of soil organic matter, resulting from the balance between the rate of accumulation of plant and animal residues and the rate of their decomposition. A virgin soil usually contains a mixture of fresh and old organic matter. The older the residue, the more thorough is the degree of decomposition. The most persistent product of organic matter decomposition is called *humus*. It consists of a mixture of more or less stable organic compounds of high molecular mass, polysaccharides and plyuronides, as well as aromatic and aliphatic compounds and amino acids. Humus exists in colloidal form, which attaches itself by adsorption to the mineral particles of the soil and serves to glue those particles in composite associations called *aggregates*. Due largely to the presence of carboxylic (–COOH) and phenolic (–OH) groups, humus tends to exhibit a negative electrostatic charge that is pH-dependent and is capable of adsorbing and exchanging cations in a manner somewhat similar to that of clay particles. Humus also tends to adsorb various pollutants such as pesticides. It reacts with metal ions to form complexes that can help mitigate the toxicity of such elements as lead and zinc. Finally, and very importantly, organic matter serves as a source of such essential plant nutrients as nitrogen, phosphorus, and sulfur.

Humus is generally dark in color, and even small amounts can impart to the soil a grayish hue. The mass ratio of carbon to nitrogen (C/N) varies from 1:10 to 1:15, and the ratio of carbon to phosphorus is about 1:100 but varies from soil to soil. The nutrients in humus are released slowly in the soil.

BOX 10.1 Soil Organic Matter

The term *soil organic matter* (SOM) consists of carbon-based compounds that result from the partial decay of plant and animal residues in the soil. It includes all forms of organic materials in soils, including litter, light fraction, microbial biomass, water-soluble organics, and stabilized organic matter. The relatively stable fraction of SOM is generally comprised of organic compounds that have a turnover rate, or mean residence time, greater than 5 years. The most recalcitrant fraction of SOM may have a mean residence time of centuries under natural conditions. That stable fraction is called *humus*. Its residence time is greatly shortened by the mechanical manipulation and pulverization of the soil, i.e. by tillage.

The amount and distribution of organic carbon in agricultural soils is strongly influenced by the crop or crops grown, and by the mode of tillage. Generally, the more intensive and prolonged the tillage regimen, the greater its effect in reducing the residual content of soil organic matter, particularly in the topsoil layer.

Converting atmospheric CO_2 into stable organic carbon in the soil can be enhanced by a set of practices, including the growing of dense crops with a high residue content (with fertilization and water management to promote growth) and the avoidance of nonessential manipulation of the soil; i.e., adopting minimal or zero tillage.

In the past, human-managed terrestrial systems were a major source of CO_2 and other GH gases emitted to the atmosphere. The trend in recent years has been to reduce those emissions and even to reverse the process from net emission to net absorption of carbon into vegetation and its sequestration in the soil.

Though humus appears to be stable relative to fresh organic residues such as rapidly decomposable plant remains and animal manures, it is not entirely so. Starting with the burning of vegetation during the Paleolithic period, continuing with the advent of agriculture during the Neolithic period, and greatly accelerating with the Industrial Revolution's use of fossil fuels, human interference has disrupted the natural carbon balance between the terrestrial domain and the atmosphere. Where soils are cultivated repeatedly and the crops are removed without replenishment, the continuing processes of microbial decomposition eventually cause the breakdown and depletion of humus itself. Where this happens, the soil typically loses fertility and tends to undergo structural deterioration (breakdown of aggregates, formation of crust, and erosion by water and

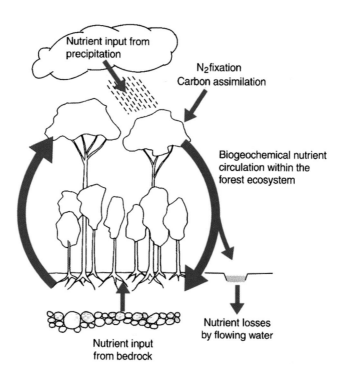

Fig. 10.1. Nutrient supply in a forest ecosystem. (Schultz, 2005).

BOX 10.2   CO2 Exchange between Soil and Atmosphere

The soil plays a vital role in the global exchange of gases and thus influences the composition of the atmosphere. A case in point is the cycling of carbon, a major constituent of living organisms. Green plants absorb carbon dioxide from the atmosphere in the process of photosynthesis, by which they produce carbohydrates. Though the CO2 content of the air has risen by some 30% in the last two centuries (since the Industrial Revolution), it still constitutes less than 0.04% of the total atmosphere. However, this seemingly small fraction is the key to life. The soil accumulates the residues of the organically bound carbon due to the growth and decay of terrestrial plants and animals.

Not all the carbon accumulated in the soil is readily recycled. The decay-resistant fraction of it remains in the soil in a quasi-stable form called *humus* (the Latin term that originally meant "soil"). Humus is now generally defined as the form of organic matter in the soil that helps to stabilize soil aggregates and to retain nutrients, while immobilizing potentially toxic pollutants. The quantity of organic carbon stored in the world's soils and in the vegetation they support is greater than the quantity present in the atmosphere as CO2. (Still more carbon is contained in mineral form in carbonate minerals, such as calcite and dolomite, often present in soils as well as in bedrock.)

In a well aerated soil, aerobic microorganisms normally oxidize organic carbon to carbon dioxide, which is then returned to the atmosphere. In a soil that is not well aerated (e.g., if it is waterlogged), anaerobic microorganisms emit methane (as well as nitrous oxide) to the atmosphere. In either case, the net release of these radiatively active gases from the soil may contribute significantly to the atmospheric greenhouse effect.

The increase of CO2 content of the atmosphere may enhance the photosynthetic productivity of green plants. (Greenhouse operators employ the same principle to stimulate crop growth by enriching the air inside their enclosures with carbon dioxide.) However, that increase is a subject of great concern at present because of its potential contribution to global warming. On the other hand, if properly managed, the soil can serve as a net sink (rather than source) of carbon dioxide, and thus help to mitigate, rather than to exacerbate, global warming. We shall elucidate the topic of carbon sequestration in soils more fully in a subsequent section of this book.

wind). Currently, an estimated 220 gigatons of carbon are released to the atmosphere from the world's soils and only some 215 are absorbed. That imbalance is a cause of the current concern over global warming. Maintenance of soil organic matter, particularly of humus, is a hallmark of proper soil management.

SOIL ACIDITY AND pH

The chemical attributes of a soil include the composition, state, and properties of its mineral and organic constituents, as well as the kinds, rates, and products of their reactions over time in different circumstances. One of the most important properties of a soil is its degree of acidity, or its opposite, alkalinity.

Strictly speaking, acidity is not an attribute of the soil's solid phase (i.e., its particles) *per se*, but of the soil solution. However, it is the interaction of the solid phase, especially the adsorbed and exchangeable ions of the clay fraction, with the water present in the soil that imparts to the solution its quality of acidity or alkalinity. In turn, that attribute of the solution governs such important phenomena as ion mobility, dissolution and precipitation of electrolytes, and oxidation-reduction reactions. The acidity of the soil solution also affects the availability of nutrients to plants and the suitability of the soil to serve as a medium for a labile biotic community.

Acidity of an aqueous solution is due to the presence and concentration of the hydrogen ion, H^+. Because of its small mass and the tightness with which its single electron is bound to the oxygen atom, the nucleus of the hydrogen atom in the water molecule exhibits a finite tendency to dissociate from the oxygen with which it is covalently associated and to "jump" to the adjacent water molecule, to which it is hydrogen-bonded. Such an event produces two ions, the hydronium ion (H_3O^+) and the hydroxyl ion (OH^-). The reaction described is reversible and should be written as $2H_2O \rightleftharpoons (H_3O)^+ + OH^-$. However, by convention it is written simply as

$$H_2O \rightleftharpoons H^+ + OH^-$$

and one speaks of "hydrogen ions" rather than of "hydronium ions".

Although the number of water molecules undergoing ionization at any given time is very small (10^{-7} mol/L) relative to the total number of water moles present per liter ($1000/18 = 55.5$ mol/L), the consequences of self-dissociation are very important. The ions' product of water is the product of the concentrations of H^+ and OH^-, namely 10^{-14}. If either the hydroxyl ion concentration [OH^-] or the hydrogen ion concentration [H^+] is changed, the companion ion concentration changes automatically to maintain the constancy of the product.

An excess concentration of hydrogen ions over the corresponding concentration of hydroxyl ions imparts to the aqueous medium the property of *acidity*, whereas a predominance of hydroxyl ions produces the opposite property of *alkalinity* or *basicity*. A condition in which the two concentrations are equal is termed *neutrality*. The pH scale is a means of designating the concentration of hydrogen ions, and hence also of hydroxyl ions. It is defined as:

$$pH = \log_{10} 1/[H^+] = -\log_{10}[H^+]$$

The pH of a neutral medium is 7, that of an acidic medium is less than 7, and that of an alkaline solution is greater than 7. Note that, due to the logarithmic nature of the pH scale, a unit change in pH implies a tenfold change in the hydrogen ion concentration.

In the soil, the pH may vary widely, from about 2 for very acidic soils to 9 or more in alkaline (sodic) soils. Highly leached, humid-region soils tend to be acidic, especially if the parent material is derived from such acidic rocks as granite. Organic soils, such as those formed in marshy conditions (e.g., peat and muck) also tend to be acidic. On the other hand, desert-region soils tend to acquire an alkaline reaction due to the presence of calcium ions, and, in places, the presence of sodium ions. Those cations also affect soil structure, and thus pH is also related to physical, as well as chemical, soil conditions.

Rainwater is naturally slightly acidic (pH 5.6) and tends naturally to cause gradual soil acidification. The natural acidity of rain is exacerbated by industrial emissions of gaseous sulfur dioxide and other acidifying agents that mix with the rain over large areas; e.g., in the northeastern U.S. and western Europe. The effect of acidification is to solubilize and leach out some nutrients (such as calcium and magnesium) and to increase the concentration of potentially toxic metal ions (such as Al, Mn, Cr, Cu, Ni, and Zn). Soil acidity also inhibits the activity of some soil microorganisms, including some of the symbiotic nitrogen-fixing bacteria. The natural or induced acidity of soils can be countered by the addition of lime, $CaCO_3$. The applied calcium serves to replace the hydrogen ion in the soil's exchange complex.

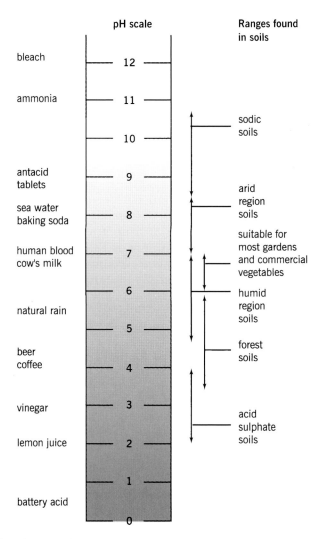

Fig. 10.2.   The pH scale and its variation in the soil. (Dubbin, 2001).

OXIDATION-REDUCTION PROCESSES

Well-aggregated, well-aerated soils generally promote the oxidation of minerals and organic matter, as well as the unrestricted aerobic respiration of plant roots and the host of other organisms living in the soil. In contrast, waterlogged and compacted soils are likely to exhibit anaerobic conditions, which induce a series of chemical and biochemical reduction reactions. Among such reactions are *denitrification*, by which nitrate is reduced to nitrite, then to nitrous oxide, and finally to elemental nitrogen: $NO_3^- \rightarrow NO_2^- \rightarrow N_2O \rightarrow N_2$. Other processes likely to occur in anaerobic soils are the reduction of manganese from the manganic to the manganous state, of iron from the ferric to the ferrous state, and sulfate reduction to hydrogen sulfide. Various organic compounds are produced in the anaerobic decomposition of organic matter. Some of the numerous products of these processes are toxic to plants; e.g., hydrogen sulfide, methane, and ethylene, as well as acetic, butyric, and phenolic acids.

Soil organisms obtain the energy they need for their vital processes from a series of chemical reactions involving the transfer of electrons from substances that serve as sources of energy to substances that are products of respiration. In aerobic respiration, the final electron sink is oxygen, which accepts electrons and combines with hydrogen ions to form water. Where free oxygen is scarce, several other substances will undergo chemical reduction. Some variable-valency ions will accept electrons and thus be reduced to a lower valency state. Examples of the reduction of high-valency iron and manganese ions:

$$Fe^{3+} + e^- \rightarrow Fe^{2+}$$
$$Mn^{4+} + 2e^- \rightarrow Mn^{2+}$$

The tendency of a solution to donate electrons to a reducible substance or to accept electrons from an oxidizable substance is generally characterized in terms of its oxidation-reduction potential, commonly known as *redox potential*. It is defined as the potential in volts required in an electric cell to induce oxidation at the anode and reduction at the cathode. Measurements of the redox potential of the soil solution can indicate the aeration state of a soil; i.e., whether the soil medium is in an oxidized (well-aerated) or a reduced (poorly aerated) state.

Poorly aerated soils often occur in humid regions characterized by high year-round rainfall. They may also occur in semihumid or even in arid regions, in low-lying river valleys, deltas and estuaries, and along marshy seacoasts. They can generally be recognized by their waterlogged condition, although occasionally a poorly aerated soil may appear to be dry on the surface. A pit dug into such a soil will often emit such gases as hydrogen sulfide, noted for its foul smell, and methane, which is highly flammable. Such gases result from the anaerobic decomposition of organic matter.

Prolonged anaerobic conditions also cause the chemical reduction of mineral elements (e.g., iron and manganese, as shown above), as a consequence of which the soil profile may acquire a bleached, grayish (occasionally even a bluish or greenish) hue. A clear telltale sign of poor aeration is a condition known as *mottling*, characterized by the appearance of discolored spots at some depth in the soil, typically just above the water table. These spots generally indicate that some of the normally reddish ferric oxides have been reduced to ferrous oxides. In extreme cases, manganese is reduced from an oxide to the elemental state and occurs in the form of small blackish nodules. Soils with such symptoms of poor aeration, typically resulting from the blockage of pores by excessive wetness, are called *gley soils* in the classical pedological terminology, or, more simply, *hydromorphic soils*.

Even apparently well-aerated soils may contain zones that are poorly aerated. In some cases, they occur as isolated spots throughout the profile; for example, in the interiors of aggregates, where water is retained for long periods. Pores between aggregates are generally well aerated because water quickly drains out of them, so they become air-filled even at a slight suction. Consequently, the outer portion of an aggregate may be bright red and seem to be completely oxidized, while the interior of the same aggregate may be mottled. Plant roots may grow all around the periphery of such aggregates but seldom penetrate into them.

BOX 10.3 Oxidation and Reduction

By definition, *oxidation and reduction* as used in chemistry refer to reactions involving the transfer of one or more electrons from one element or compound to another. The entity losing electrons is said to undergo *oxidation*, whereas the entity gaining electrons is said to undergo *reduction*.

Consider, for a simple example, the combination of elemental hydrogen (H_2) with elemental oxygen (O_2) to form water (H_2O). Initially, both elements are regarded as neutral (neither positively nor negatively charged). When they combine to form H_2O, in effect each of the hydrogen atoms acquires a positive charge, having given its electron to an oxygen atom, which thereby gains two electrons. In other words, the hydrogen is oxidized while the oxygen is reduced. Thus:

$$O_2 + 4H^+ + 4e^- \rightarrow 2H_2O$$

Many oxidation processes, as the name implies, involve the combination of an element with oxygen; for instance, in the oxidation of sugar to form carbon dioxide and water:

$$C_6H_{12}O_6 + 6O_2 \rightarrow 6CO_2 + 6H_2O$$

Despite the name given to the reaction, however, many oxidation processes involve substances other than oxygen as electron acceptors. Consider, for example, the reduction of nitrate to nitrite, followed by the reduction of nitrite to elemental nitrogen:

$$NO_3^- + 2H^+ + 2e^- \rightarrow NO_2^- + H_2O$$

$$2NO_2^- + 8H^+ + 6e^- \rightarrow N_2 + 4H_2O$$

The opposite process is the oxidation of ammonium to nitrite, and, in turn, of nitrite to nitrate.

Another example is the reduction of iron from the trivalent state (Fe_2^+) to the divalent (Fe_2^+) state:

$$Fe(OH)_3 + 3H^+ + e^- \rightarrow Fe^{2+} + 3H_2O$$

Oxidation-reduction processes are mediated by soil microorganisms and often involve organic (carbon-containing) compounds of biological origin. The complete oxidation of carbon in organic matter results in the emission of carbon dioxide gas (CO_2). The complete reduction of carbon in the soil, generally in the absence of oxygen (i.e., in anaerobic conditions), commonly results in the emission of methane (CH_4). Both, as it happens, are radiatively active gases (so-called "greenhouse gases") when released into the atmosphere. So, incidentally, is the gas nitrous oxide (N_2O).

SOIL SALINITY AND SODICITY

The term *soil salinity* refers to the presence of electrolytic mineral solutes in concentrations that are harmful to many plants in the soil and in the aqueous solution within it.[1] Most common among these solutes are the dissociated cations Na^+, K^+, Ca^{2+}, and Mg^{2+}; along with the anions Cl^-, SO_4^{2-}, NO_3^-, HCO_3^-, and CO_3^{2-}. In addition, the soil solution in hypersaline soils may contain trace concentrations of the elements B, Se, Sr, Li, Si, Rb, F, Mo, Mn, Ba, and Al, some of which may be toxic to plants and animals.

Some salts may result from the original chemical decomposition (weathering) of the minerals that constitute the rocks from which the soil was derived. Other quantities of salt may enter the soil with rainwater. Although the initial condensation of vapor produces pure water, the raindrops that form in clouds and fall earthward tend to pick up soluble constituents during their brief residence in the atmosphere. One such constituent is carbon dioxide, which dissolves in rainwater to form a dilute solution of carbonic acid. That acid, though relatively weak, reacts with minerals in dust, rocks, and the soil, and causes certain minerals to dissolve more readily than they would otherwise, thus contributing indirectly to soil salinity. The acidity of rainwater rises significantly in industrialized regions where it mixes with emitted gases such as oxides of sulfur and nitrogen, forming a dilute solution of sulfuric acid and nitric acid, known as "acid rain". In addition, rainfall that occurs in coastal regions often

---

[1] Plants that are especially adapted to grow under saline conditions are called *halophytes*. Plants that are not so adapted, called *glycophytes*, generally exhibit symptoms of physiological stress when subjected to salinity. Most crop plants are of the latter category, though they differ from one another in the degree of their sensitivity.

mixes with sea spray, which may contribute appreciable quantities of salt to areas that extend some distance from the shore. Seawater tends to intrude landward via tidal estuaries, as well as into groundwater aquifers.

Except along seashores, however, saline soils seldom occur in humid regions, thanks to the net downward percolation of fresh water through the soil profile, brought about by the excess of rainfall over evapotranspiration. Arid regions, in contrast, may have periods of no net downward percolation and hence no effective leaching, so salts can accumulate in the soil. Hence, the combined effects of meager rainfall, high evaporation, the presence of salt-bearing sediments, and, particularly in river valleys and other low-lying areas, the occurrence of shallow, brackish groundwater, give rise to saline soils.

BOX 10.4   Sample Calculation of the Salt Balance

The following data were obtained in a field in an arid zone:

1. Rainfall occurred only in winter and amounted to 300 mm, with a total salt concentration of 40 ppm.
2. Capillary rise from saline groundwater in spring and autumn totaled 100 mm at a concentration of 1000 ppm.
3. Irrigation was applied during the summer season and amounted to 900 mm, with 400 ppm salts.
4. Drainage during the irrigation season amounted to 200 mm, with a soluble salt concentration of 800 ppm.

An additional increment 0.12 kg/m² of soluble salts was added in the form of fertilizers and soil amendments, of which 0.1 kg/m² was removed by the harvested crops.

Disregarding possible dissolution of soil minerals and the precipitation of salts within the soil, compute the annual salt balance. Is there a net accumulation or release of salts by the soil?

We begin with the salt balance equation of the root zone per unit of land area:

$$\Delta M_s = \rho_w (V_r c_r + V_i c_i + V_g c_g - V_d c_d) + M_a - M_c$$

Herein ΔM_s is the change in mass of salt in the root; ρ_w is the density of water; V_r, V_i, V_g, V_d are the volumes of rain, irrigation, groundwater rise, and drainage, respectively, with corresponding salt concentrations of c_r, c_i, c_g, c_d; and M_a, M_c are the masses of salts added agriculturally and removed by the crops, respectively.

Using 1 m² as the unit "field" area, calculating water volumes in terms of m³ and masses in terms of kg (water density being 1000 kg/m³), we can substitute the given quantities into the previous equation to obtain the change in mass content of salt in the soil:

$$\Delta M_s = 10^3 \text{kg/m}^3 \ (0.3\text{m})(40 \times 10^{-6}) + (0.1\text{m})$$
$$(1000 \times 10^{-6}) + (0.9\text{m})(400 \times 10^{-6}) - (0.2\text{m})$$
$$(800 \times 10^{-6}) + 0.12 \text{kg/m}^2 - 0.1 \text{kg/m}^2$$

$$\Delta M_s = 0.322 \text{ kg/m}^2\text{yr}$$

The root zone is accumulating salt at a rate of 3220 kg (3.22 metric tons) per hectare per year.

Less obvious than the appearance of naturally saline soils, but perhaps more insidious, is the unintended *induced salination* of originally productive soils, caused by human intervention. Irrigation waters generally contain appreciable concentrations of salts. Crop plants normally extract water from the soil while leaving most of the salt behind. Unless leached away by excess percolation, such salts tend to accumulate in the root zone. Where internal drainage is impeded (e.g., by the presence of an impervious layer in the subsoil), attempts to leach the soil can do more harm than good, by raising the water table, waterlogging the soil, and causing capillary rise to the surface, where evaporation infuses the topsoil with cumulative quantities of residual salts.

Overall salinity is usually expressed in terms of *total dissolved solutes* (TDS) in milligrams per liter of solution (approximately equivalent to parts per million, ppm). Salinity may also be characterized by measuring the *electrical conductivity* (EC) of the solution, generally expressed in terms of decisiemens per meter (dS/m). Quantitative criteria for diagnosing soil salinity were originally formulated by the US Salinity Laboratory at Riverside, California (in a handbook first issued in 1954) in terms of the EC of the soil's saturation extract; i.e., the solution extracted from a soil sample that had been presaturated with water. Those criteria are still in wide use today. Accordingly, a saline soil has been defined as having an EC of more than 4 dS/m. This value generally corresponds to approximately 40 mmol of salts per liter. In the case of a sodium chloride solution (NaCl being the most commonly occurring salt), this criterion corresponds to a concentration of about 2.4 grams per liter.

The clay fraction in saline soils is generally well flocculated. As the salts are leached, however, the flocs tend to disperse and the soil aggregates to break down. This is especially likely to occur where the sodium ion predominates in the exchange complex of the soil. The dispersion of clay may result in deterioration of soil structure by clogging of large pores in the soil, and consequently in the reduction of the soil's permeability to water and air. This phenomenon is generally referred to as *soil sodicity*, or *alkalinity*. When wet, the soil's top layer becomes a slick and sticky mud; when dry, it hardens to form a tough crust with an irregular, roughly hexagonal, pattern of cracks. This dense surface condition forms a barrier to the emergence of seedlings and to the penetration of their roots. As infiltration is inhibited, greater runoff, erosion, and silting of downstream water courses and reservoirs ensues.

Table 10.1   Classification of Water Quality According to Total Salt Concentration

Designation	Total dissolved salts (ppm)	EC(dS/m)	Category
Fresh water	<500	<0.6	Drinking and irrigation
Slightly brackish	500–1,000	0.6–1.5	Irrigation
Brackish	1,000–2,000	1.5–3	Irrigation with caution
Moderately saline	2,000–5,000	3–8	Primary drainage
Saline	5,000–10,000	8–15	Secondary drainage and saline groundwater
Highly saline	10,000–35,000	15–45	Very saline groundwater
Brine	>35,000	>45	Seawater

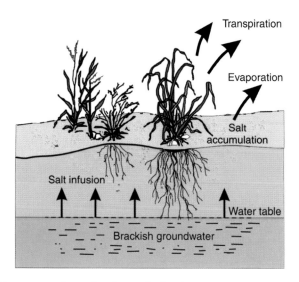

Fig. 10.3.   The process of waterlogging and salination.

BOX 10.5 Sodium Adsorption Ratio (SAR) and Exchangeable Sodium Percentage (ESP)

A widely accepted index for characterizing the soil solution with respect to its likely influence on the exchangeable sodium percentage in the *sodium adsorption ratio* (SAR). It has also been used to assess the quality of irrigation water. SAR is defined as follows:

$$SAR = [Na^+]/\{([Ca^{2+}] + [Mg^{2+}])/2\}^{0.5}$$

In words, SAR is the ratio of the sodium ion concentration to the square root of the average concentration of the divalent calcium and magnesium ions. In this context, all concentrations are expressed in milliequivqalents per liter. SAR is thus an approximate expression for the relative activity of Na^+ ions in exchange reactions in soils. A high SAR, particularly at low concentrations of the soil solution, causes high ESP and is also likely to cause a decrease of soil permeability.

The relationship between ESP and SAR of the soil solution was measured on numerous soil samples in the Western states by the U.S. Salinity Laboratory and reported (Richards, 1954) to be

$$ESP = 100[-a + b(SAR)]/\{1 + [-a + b(SAR)]\}$$

where $a = 0.0126$ and $b = 0.01475$.

The tendency toward swelling and dispersion is highest in soils with a high content of smectite clay (an active clay also named montmorillonite) and when sodium ions are a sizable fraction of the cations exchange complex. In practice, the criterion for a "sodic soil" is an *exchangeable sodium percentage* (ESP) of 15 percent or more of the soil's total cation exchange capacity. This is, admittedly, a rather arbitrary criterion, since in many cases no sharp distinction is apparent at a particular value of ESP between "sodic" and "non-sodic" soil conditions.

In principle, a soil may be saline without being sodic, and in other circumstances, a soil may be sodic without being saline. All too often, however, irrigated soils in arid regions can be both saline and sodic; i.e., when the EC exceeds 4 dS/m and ESP exceeds 15%. Such soils, when leached, tend to become strongly alkaline, with their pH values rising above 8.5. Sodic soils may appear darkish, the coloration due to the surface coating of dispersed organic matter. For this reason, sodic soils were long ago described as "black alkali", in contrast with non-sodic saline soils that were described as "white alkali" because of the typical appearance of crystalline salt at the surface.

Saline soils can, in principle, be ameliorated by leaching with fresh water. However, this can only be accomplished if the soil is well drained. Drainage

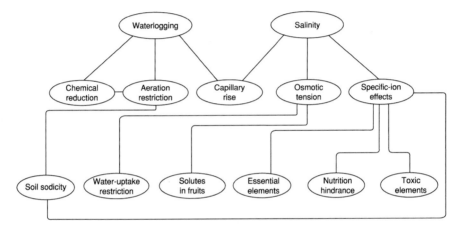

Fig. 10.4. Effects of salinity and sodicity on soils and plants.

Fig. 10.5. Global distribution of salt-affected soils.

Table 10.2   Global Distribution of Salt-Affected Soils (Szabolcs, 1979)

Continent	Area in millions of ha		
	Saline	**Sodic (alkali)**	**Total**
North America	6.2	9.6	15.8
Central America	2.0	—	2.0
South America	69.4	59.6	129.0
Africa	53.5	27.0	80.5
South Asia	83.3	1.8	85.1
North and Central Asia	91.6	120.1	211.7
Southeast Asia	20.0	—	20.0
Australasia	17.4	340.0	357.4
Europe	7.8	22.9	30.7
Total	351.5	581.0	932.2

BOX 10.6   Saline Seeps in Australia and North America

Human management of the land may lead to unforeseen environmental con-
secuences. A striking example is the saline seep phenomenon in Australia and
North America. A few decades ago, Australian farmers were startled to notice
that brine began to ooze out of low spots in their fields. Subsequent research revealed
the phenomenon to be the delayed result of the extensive clearing of land carried out in
southern Australia a century earlier.

Rains falling on land along the coasts of Australia mix with the sea spray, which forms
by the splash of ocean waves against the shore and is wafted onto the land by the winds.
Over eons of history, the brackish rains deposited a load of salt onto the soil. Native forests extracted moisture from the soil but left the salt behind. Slowly, the saline solution
percolated beyond the root zone, gradually charging the subsoil with residual salts.

When European settlement of Australia began in the mid-1800s, the settlers cleared
away the forest along the southeastern and southwestern edges of the continent. This
changed the water regime. In the natural evergreen forest, part of the rain had been intercepted by the foliage and then evaporated without ever reaching the ground, and moisture
extraction by the deep roots was continuous. The seasonal, shallow-rooted crops planted
by the settlers extracted and evaporated less moisture. The increased fraction of rainfall that
percolated downward eventually came to rest over an impervious stratum and formed a
water table. That water also mobilized the salt that had accumulated in the subsoil in eons
past. Gradually, the water table rose, and with it returned the ancient salt. All the while,
the farmers and ranchers were oblivious to what was happening down below, so when the
brine reached the surface and formed splotches of sterile soil, it came as a total surprise.

Saline seeps also occur in North America, both in the U.S and in the prairies of Canada. Tractors bog down or cut deep ruts in the low spots where the brine flows out. Here,
the original vegetation was perennial grass, not trees, and the source of the salts was
not sea spray but underlying deposits of marine shales, originating in an earlier geologic
era. However, the process by which the seeps became manifest was similar in both
continents: it resulted from human disruption of a preexisting equilibrium.

The saline seep problem can be remediated in many locations, but the task may well require time and considerable expense. Seasonal, shallow-rooted crops can be replaced with evergreen, perennial, deep-rooted vegetation capable of lowering the water table over the entire area contributing to the seeps, and drains can be installed within the seeps themselves to ensure the environmentally safe disposal of the emerging brine.

of waterlogged and salinity-prone soils is done by means of open ditches or embedded perforated pipes, which must be deep enough to reach beneath the water table. The drainage water must then be disposed of safely by discharging it into the sea, if the coast is near enough, or into closed evaporation ponds. The common practice of discharging brackish effluent into a river is likely to result in the salination of the river itself, which can harm downstream users as well as the entire natural ecology of the river.

In practice, salts are more difficult to leach from clayey than from sandy soils, because of the lower hydraulic conductivity and higher ionic retentivity of clay, and because of its tendency to disperse in the presence of exchangeable sodium. To prevent the hazard of sodicity in a soil leached of excess salts, a common practice is to add a *soil amendment*, generally consisting of substances containing a divalent cation such as calcium, which promotes clay flocculation and thus counters the tendency of sodium to cause dispersion of the clay. The most prevalent soil amendments for the purpose of improving soil structure and permeability are dihydrate calcium sulfate (gypsum) and dihydrate calcium chloride. Of these, gypsum is generally preferred due to its ready availability in many places and to its relatively low cost. It may be derived from mining or be available as a by-product of the phosphate fertilizer industry. The amount of gypsum needed to replace the exchangeable sodium depends on the initial exchangeable sodium percentage, the soil's total cation exchange capacity, its bulk density, and the depth of the soil to be treated effectively.

BOX 10.7 Salination and Alkalization

Soil salination is the process whereby salts—generally chlorides or sulfates—accumulate. The salts may originate in sea spray (especially along coastlines), in brackish irrigation waters, or may be produced by weathering of minerals. Such salts accumulate whenever the rate of evaporation exceeds the rate of percolation. Soil salinity reduces crop productivity, and may, in extreme cases, render the soil practically sterile to all but specialized salt-loving plants (known as *halophytes*).

Alkalization is the process by which the exchange complex of the soil accumulates sodium ions that may replace the cations calcium and magnesium. When the proportion of sodium ions in the exchange complex exceeds 15% or so, the pH rises above 8, clay tends to disperse, and the aggregated structure of the soil then collapses. The soil then becomes swollen, slick, and anaerobic when wet, but dense and hard when dry.

11. SOIL FERTILITY AND PLANT NUTRITION

Whoever could make two ears of corn to grow
upon a spot of ground where only one grew before
would deserve better of mankind
than the whole race of politicians put together.
Jonathan Swift, 1667–1745

SOIL AS A MEDIUM FOR PLANT GROWTH

The capacity of a soil to serve as a favorable medium for plant growth depends on several interrelated attributes. The soil must be porous and permeable enough to permit the free entry, retention, and transmission of water and air. It must also contain a supply of nutrients in forms that are available to plants but that do not leach too rapidly. The soil should be deep and loose enough to allow the roots to penetrate and proliferate. In addition, the soil must have an optimal range of temperature and pH, and be free of excess salts or of toxic factors. The most productive soils are on level or slightly sloping terrain in the midlatitudes, with an adequate but not excessive supply of water, good drainage and aeration, a sufficient supply of nutrients and effective protection against erosion. However, some plant communities and some crops—notably rice and sugarcane—can grow well in flooded soils that are poorly aerated.

MINERAL NUTRITION OF PLANTS

The essential function of green plants is *photosynthesis*, by which carbon dioxide absorbed from the atmosphere is combined with water derived from

151

the soil to synthesize the basic materials of life. That function is powered by the infusion of radiant energy from the sun. The simple sugars initially synthesized are later subsequently elaborated by the plants to create the many complex chemical compounds that serve the myriad forms of life dependent on the soil. In the process, plants absorb various elements from the soil, some 18 of which are considered essential nutrients for plant growth. The ability to supply those elements to plants in optimal quantities to promote their growth determines the soil's *chemical fertility*.

Of the 18 elements known to be required for plant growth, nine are needed in relatively large quantities and are therefore called *macronutrients*, and nine are needed in relatively small quantities and are called *micronutrients* (or *trace elements*). The macronutrients include carbon (C), hydrogen (H), oxygen (O), nitrogen (N), phosphorus (P), potassium (K), sulfur (S), calcium (Ca), and magnesium (Mg). The micronutrients include iron (Fe), manganese (Mn), boron (B), zinc (Zn), copper (Cu), chlorine (Cl), molybdenum (Mo), cobalt (Co), and nickel (Ni). Other elements are taken up by plants and may be needed by some of them but are not considered generally essential.

Of the essential elements, carbon, oxygen, and hydrogen are obtained by a plant from the atmosphere and from water, whereas all the other essential elements are derived from the soil. To be available to plants, these elements must be present in the soil solution in sufficient concentration, in regions of the soil that are accessible to plant roots. Nutrients tend to move toward the roots by diffusion in response to concentration gradients in the water phase surrounding the roots, as well as by convection during the uptake of soil-water by the roots. At the same time, the roots of growing plants also move, by extension and proliferation, toward untapped nutrient-rich regions within the soil.

In principle, plant nutrients may be present in the soil in three possible forms: (1) *readily bioavailable*—present in the soil solution within the active rooting zone of the plants requiring them; (2) *labile reserve*—adsorbed in the soil's exchange complex and/or present in rapidly decomposable organic matter; and (3) *stable reserve*—present in relatively unweathered soil minerals and in slowly decomposable organic compounds (humus). Often, the quantity of readily bioavailable nutrients is only a small fraction of the potentially available quantity held in reserve in exchangeable or insoluble forms. In the latter case, the rate of release of the reserve nutrients is a principal determinant of the soil's inherent fertility. Another determinant is the physicochemical condition of the soil: it must be such as to allow proper root function; e.g., the soil must not be too dry, acidic, saline, compact, anaerobic, or cold.

In natural ecosystems, the nutrients taken up by plants from the soil are recycled as plant products, and the products of animals that consume the plants, are returned to the soil and decomposed in it. In agricultural production, however, some of the plant products are removed. Nutrients originally available in the soil are thus extracted and may become deficient. Where one or another of the essential nutrient elements is leached, depleted, or tied up physicochemically, the plants are likely to suffer from impaired physiological function. Hence, to be sustainable, agricultural production typically requires artificial fertilization.

For each nutrient and plant type, there is an optimal range of distribution and concentration below which and above which plant growth is constrained.

While some essential nutrients may be available in sufficient amounts, one or another of the nutrients may be insufficient and hence constitute the limited factor in the soil. Optimal plant growth requires an appropriate balance of all essential factors. The soil is, in sum, a dynamic reservoir of water, nutrients, and chemical energy. Its capacity to store, buffer, and release vital nutrients to living organisms determines the quality of the soil and its role in the biosphere.

BOX 11.1 Nutrient Uptake

Soluble nutrients in the soil solution reach the roots both by diffusion due to concentration gradients and by convection, or mass flow. The latter is impelled by the *transpiration stream*, a catenary transport process in which water is extracted by the atmosphere from the leaves, which in turn draw water via the plant's vascular system from the roots. The roots, for their part, grow and proliferate in the soil in a constant quest for water and nutrients.

In some cases, roots may absorb ionic nutrients by direct contact with the ion-exchange complex on the surfaces of clay particles and of organic matter. The presence of symbiotic fungi known as *mycorrhizas* may facilitate such absorption by mediating between the roots and the soil, thus facilitating the uptake of such nutrients as N, P, K, Ca, Mg, Cu, and Zn. Nitrogen is generally absorbed by plant roots from the soil solution in the forms of NO_3^- and NH_4^+. Phosphorus is absorbed in the forms of $H_2PO_4^-$ and HPO_4^-, and sulfur in the form of SO_4^-. The nutrient metals are absorbed in the form of the cations K^+, Ca^{++}, and Mg^{++}.

In stable natural ecosystems, the additions and losses of nutrients in the root zone are roughly in balance. The additions may occur by rainfall input, release from weathering of minerals, and fixation by microorganisms. The losses occur by erosion, leaching, and volatilization. In agricultural systems, where the crops are removed annually, the losses may greatly exceed the gains, so nutrients are augmented by fertilization.

PRINCIPAL MACRONUTRIENTS

The three principal macronutrients, so called because they are required in the largest quantities, are nitrogen, phosphorus, and potassium (N, P, and K). The most common fertilizers in current use are mixtures of compounds containing the three components, conventionally expressed in terms of the respective percentages of N, P_2O_5, and K_2O.

Nitrogen

First and foremost among the nutrient elements is nitrogen. Curiously, although this element constitutes nearly 79% of the atmosphere and all creatures exposed to the air live in a veritable ocean of it, the elemental form, N_2, is completely unavailable to the vast majority of green plants. This is in sharp contrast to the ability of plants to absorb CO_2 from the air, although its concentration in air is less than 0.04% (less than 400 parts-per-million). Nitrogen is a vital component of proteins, of nucleic acids that carry every species' genetic code, and of the chlorophyll that enables the process of photosynthesis and

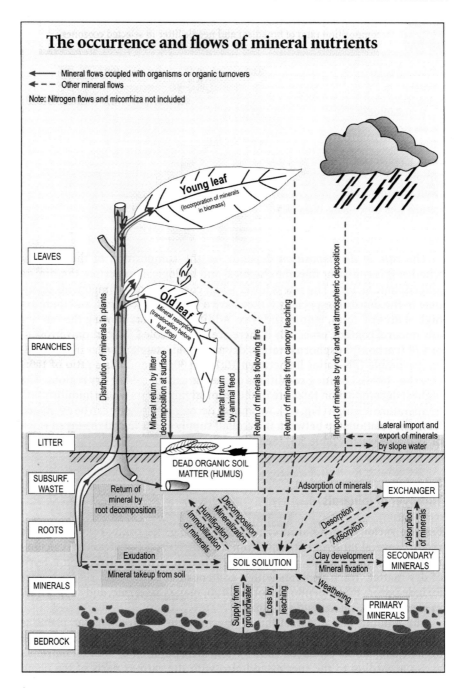

Fig. 11.1. Occurrence and flows of nutrients. (Schultz, 2005).

that imparts to the leaves of plants their typical deep green color. Plants that are deficient in nitrogen exhibit stunted growth and a pale, yellowish green hue, a condition termed *chlorosis*.

Nitrogen is not generally present in the minerals that constitute a soil's parent material, and it exists almost entirely in the soil's organic matter. In natural ecosystems, nitrogen is recycled by microbial decay from plant and animal remains within the soil or over its surface. To be available to growing plants, the nitrogen in the soil must be in soluble ("mineralized") form, either as ammonium (NH_4^+) or ,following the process of *nitrification*, as nitrate (NO_3^-). The high solubility of these forms makes them especially vulnerable to leaching by percolating water. Where such nitrates end up in groundwater or surface reservoirs, such as ponds and lakes, they may cause *eutrophication*, or the enrichment of water bodies with nutrients that result in algal blooms and decay. In waterlogged anaerobic soils, nitrogen compounds may undergo chemical reduction (*denitrification*), and may be emitted from the soil in the form of nitrous oxide gas (N_2O), which, incidentally, is a contributor to the enhanced *greenhouse effect* and global warming.

Although, as stated, green plants are not normally able to absorb atmospheric nitrogen, there are specialized microorganisms that can do so. Among the important *nitrogen fixing bacteria* are those belonging to the group known as *rhizobium*. These bacteria live in close association, called *symbiosis*, with certain species of legumes, including clover, alfalfa, vetch, peas, beans, etc. The bacteria attach themselves to the roots of these legumes, forming nodules, within which they "fix" the absorbed elemental nitrogen biochemically and transmit it to their host plants, which in turn supply the bacteria with the substances (e.g., sugars) needed for their subsistence. In agricultural practice, the seeds of leguminous plants can be inoculated, where necessary, with the appropriate bacteria so as to induce the process of symbiotic nitrogen fixation. Stands of such plants may be incorporated into the soil to enrich it with nitrogen for the purpose of supplying the needs of subsequent non-leguminous crops, a practice known as *green manuring*. Other species of free-living bacteria—among them, *cyanobacteria*—are able to fix nitrogen non-symbiotically.

Some atmospheric nitrogen is fixed and added to the soil naturally from such phenomena as lightning. An industrial process for fixing elemental nitrogen into ammonia was developed in Germany early in the twentieth century. Although that process was also used to synthesize explosives, its usefulness for peaceful purposes has been a great boon to agriculture worldwide, and contributed much to the so-called Green Revolution in the second half of the twentieth century.

Nitrogen is commonly added to the soil in organic manure or as inorganic fertilizer. The most common nitrogen fertilizers are nitrates (e.g., sodium nitrate, which occurs notably in Chile and is known as *Chile saltpeter*), or ammonia (e.g., ammonium sulfate). Ammonium nitrate is a concentrated fertilizer that combines the two chemical forms. Still another form of nitrogen used as fertilizer is urea, which is generally oxidized in the soil into nitrate. Soluble forms of nitrogen can also be sprayed in dilute solution directly onto the foliage of crops.

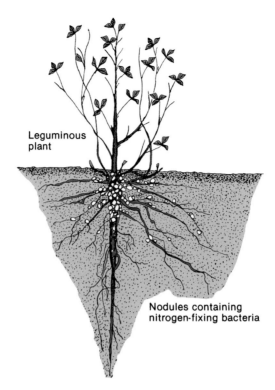

Fig. 11.2.   Nodules of nitrogen-fixing rhizobium bacteria on the roots of a legume.

Phosphorus

Phosphorus is an essential component of plant cell membranes and of nucleic acids. It is a crucial constituent of the vital molecule known as *adenosine triphosphate* (ATP), which supplies the energy for the plants' metabolic processes. Phosphorus is also an important element in animal bones and teeth. A deficiency of phosphorus reduces plant productivity and delays yield maturation. In the soil, phosphorus occurs in both organic and inorganic (mineral) forms. An important phosphorus-bearing mineral is apatite, which, however, is soluble only in acidic conditions. In mature soils that are strongly acidic, the phosphorus originally present in such minerals as apatite may have been completely dissolved and leached away, or bound into insoluble form with iron and aluminum.

To be available as a nutrient to plants, phosphorus must be present in the soil solution as the anions $H_2PO_4^{2-}$ or HPO_4^-. Most other forms are of limited solubility; hence, this element is frequently found to be deficient. The common remedy is to add a fertilizer such as superphosphate, which is obtained by treating phosphate rock—generally consisting of insoluble calcium phosphate—with sulfuric acid to obtain the more soluble forms $CaHPO_4$ or $Ca(H_2PO_4)_2$. The product of that treatment also contains calcium sulfate (gypsum), which itself can have a beneficial effect on soil structure. The added phosphorus, however,

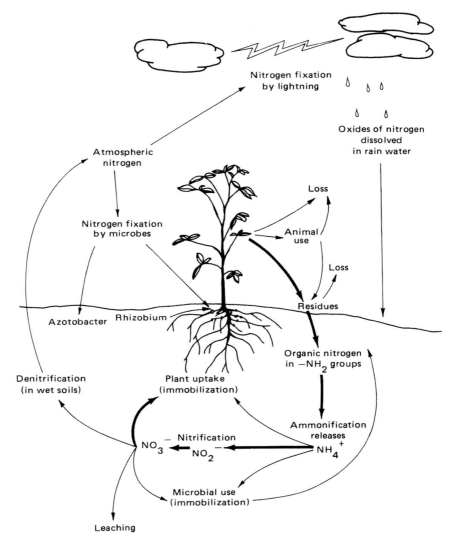

Fig. 11.3. The nitrogen cycle. (CSIRO, 1977).

may not remain in available form for long, because it tends to be converted to insoluble forms, either by reaction with calcium or with iron and aluminum oxides in the soil, or by strong adsorption onto the edges of clay crystals.

Potassium

Potassium has many and varied functions in plant life, as a constituent of enzymes and as a regulator of drought tolerance and water-use efficiency. Some crops, such as bananas and potatoes, have markedly high contents of potassium. A deficiency in this element causes chlorosis of leaf margins, and, in severe cases, the shedding of older leaves.

In the soil, the principal sources of potassium are such minerals as feldspars (particularly orthoclase) and micas, which release this element in the course of their weathering. This process makes potassium available to plants. However, where the weathering process has been prolonged and extreme, as in the cases of ultisols and oxisols, potassium may be leached away so that the soils require the addition of potassium in the form of fertilizer.

In the soil itself, potassium appears in three forms: (1) in the soil solution; (2) as the exchangeable ion K^+ adsorbed to, or released from, the surfaces of clay particles; and (3) in organic matter. Some layered-aluminosilicate clay minerals may have the capacity to adsorb potassium not only on their outer surfaces but also inside the layered crystal lattices, thereby causing the immobilization (or fixation) of potassium ions.

Other Macronutrients

Sulfur

Sulfur is a constituent of several amino acids and vitamins. Some plants—notably cabbage, onions, corn, cotton, and sorghum—have especially high requirements for sulfur. A deficiency of sulfur, which occurs most often in sandy soils and in highly weathered and leached soils, typically results in chlorosis of young leaves and in retarded plant growth.

The presence of sulfur in soils may be due to sulfide (e.g., pyrite, FeS_2) and sulfate minerals (e.g. gypsum, $CaSO_4$), the latter being particularly prevalent in arid regions. More generally, sulfur is a constituent of organic matter, from which it is released in the form of a sulfate ion by microbial action. Sulfur may also be added to the soil as fallout from atmospheric sources, including volcanic eruptions, particulate and gaseous products of burning vegetation, and gaseous emissions deposited downwind of industrial regions. Some of the atmospheric sulfur depositions may be absorbed directly by plants via their leaves, and some indirectly via their roots. In areas of concentrated atmospheric emissions of sulfur oxide, this gas mixes with rainwater to form dilute sulfuric acid, which in turn tends to acidify the soil as well. In some cases, nitrogen oxides in the atmosphere may also react with rainfall to produce nitric acid (HNO_3), which further contributes to *acid rain*. In some places (e.g., recently drained coastal wetlands), the concentration of sulfuric acid may be so high as to develop so-called *acid sulfate soils*, with pH values as low as 2.

Calcium and Magnesium

Calcium and magnesium are chemically similar—both belonging to the alkaline earth group of elements in the periodic table—and often occur together in minerals; e.g., dolomite. Plants contain 0.4 to 2.5 percent Ca and 0.1 to 0.4 percent Mg, on a dry matter basis. Calcium is essential in the development of plant buds, while magnesium is a constituent of chlorophyll molecules. Deficiencies of these elements can occur in acidic sands and leached soils, and may cause chlorosis between the veins of older leaves. Both Ca^{2+} and Mg^{2+} cations also play a role in the flocculation of colloidal clay and hence influence soil structure.

MICRONUTRIENTS

As mentioned, nine additional elements have been found to be essential nutrients for many plants, albeit in smaller quantities; hence, they are called micronutrients, or *trace elements*. These include iron (Fe), manganese (Mn), boron (B), zinc (Zn), copper (Cu), chlorine (Cl), molybdenum (Mo), cobalt (Co), and nickel (Ni). These elements play a role in various metabolic pathways, including enzyme activation. Deficiencies of micronutrients may retard the development of young plant tissues, especially of new leaves and of reproductive organs; e.g., flowers.

Micronutrients generally originate from the weathering of soil minerals. In the soil solution, they may combine with soluble humic (organic) molecules to form mineral-organic complexes known as *chelates*, which prevent the nutrients from being leached away. Such complexes occur naturally, but may also be added to the soil as synthetic amendments to increase micronutrient availability.

The availability of micronutrients in the soil solution is strongly influenced by conditions of acidity, which enhances their solubility, except in the case of molybdenum. Hence, deficiencies of most micronutrients are more likely to occur in soils of high pH. On the other hand, where the pH is especially low, several of these metals may in fact occur in the soil solution in concentrations that can become toxic to some plants.

FERTILIZERS AND SOIL AMENDMENTS

Whereas in natural ecosystems the nutrients are continually recycled, the agricultural production, harvesting, and removal of crops generally results in the net extraction of nutrients from the soil. Unless the extracted nutrients are replaced, the production system cannot be sustained. The more intensive the production system, in fact, the greater the need for the addition of nutrients becomes. The so-called Green Revolution, which greatly increased crop production in many developing countries ever since the 1960s, was achieved only when high-yielding crop varieties were developed, disseminated, and adopted in combination with intensified application of fertilizers and irrigation.

In traditional, pre-modern agriculture, the main sources of nutrients applied by farmers were plant residues and animal manures, available on the farm. Now that agricultural produce is transported to distant urban centers and fewer animals are integrated into production systems, the organic wastes are no longer available locally to be returned to the soil, and, in many cases, even become sources of pollution where they are discharged into groundwater, streams, estuaries, lakes, or seashores. Such wastes can better by applied, where possible (as sewage sludge or as dried compost), to farmland in the peripheries of cities.

Much can still be done to retain organic residues on the farm and to reincorporate them into the soil. Varying the type of crops that are grown, a practice called *crop rotation*, can also help to prevent nutrient depletion and to maintain soil fertility, especially if the cropping program includes the periodic inclusion of soil-enriching legumes. (Crop rotation also helps to prevent the proliferation of weeds, pests, and diseases in given tracts of land.) Organic materials do not contain the high concentration of nutrients that do chemical fertilizers, but, where available, they offer the advantages of slow release of nutrients (less prone to leaching) and the improvement of soil structure.

Soluble chemical fertilizers, particularly nitrogenous fertilizers, are more readily available to crops than are organic manures and composts, but are prone to leaching from the soil. Some fertilizers, such as those containing nitrate ions, may exacerbate soil acidity. Other fertilizers, such as superphosphate, may help to counter soil acidity. Where soil acidity creates conditions unfavorable for crop production, certain soil amendments can be applied to ameliorate the soil. An important and frequently used soil amendment is ground limestone (calcium carbonate), which is used to correct excessive acidity. The amount necessary depends on the degree of soil acidity (the pH) and on the soil's own buffering capacity, which in turn depends on its amount and type of clay. In the opposite case of soil alkalinity, a condition that occurs in some arid regions, the appropriate amendment is often gypsum (calcium sulfate), or even elemental sulfur, which is oxidized in the soil to produce sulfuric acid.

The transport of nutrients from the soil—either in sediments eroded from the soil surface or in the drainage waters leached from the profile—can have the effect of contaminating bodies of fresh water, including streams and lakes. The addition of nutrients, especially of nitrogen and phosphorus, in such water bodies can induce the proliferation of algae and cause originally clear waters to become turbid. Quantities of biomass may accumulate and decay in these waters, and may deplete their oxygen content. In extreme cases, this process, called *eutrophication*, can turn some freshwater lakes into foul-smelling, slimy mires.

HISTORICAL REVIEW OF SOIL FERTILITY MANAGEMENT

Traditional farming systems generally included a period of fallow in the cropping sequence to help restore soil fertility. The biblical injunction, for example, required fallowing the land every seventh year to let the land rejuvenate (Deuteronomy 15). From at least the time of Cato the Censor (234–139 B.C.E.), the Romans were also aware of the need to boost soil fertility by fallowing, as well as by crop rotation, liming acid soils, and adding manure. In medieval Europe, between one-third and one-half of the arable land was left fallow.

However, increases in population density gradually led to a reduction in the fractional area left fallow, until the custom of fallowing nearly disappeared. Spreading animal manures in the field, as well as inclusion of leguminous crops, helped to add nitrogen, a principal nutrient, to the soil. Such legumes as clover, beans, and peas can improve soil fertility because of their symbiotic association with specialized bacteria that attach themselves to plant roots and that can absorb elemental nitrogen from the atmosphere, a remarkable feat that higher plants cannot perform on their own. In the rice farming areas of Asia, the occurrence of blue-green algae (cyanobacteria) in the flooded rice paddies similarly helped to supply nitrogen to the soil, and the application of organic residues, including human wastes, further helped to maintain soil fertility.

As agricultural production was further intensified, with multiple cropping per year and with more nutrients removed from the fields as crops were harvested, extensive areas began to experience a progressive loss of soil fertility resulting from the depletion of essential nutrients. Consequently, yields began to decline. Some farmers were desperate enough to glean animal and human bones from the great battlegrounds of Europe (Waterloo, Austerlitz, etc.) in order to crush and spread them on their gardens and field plots. In 1840, Justus von Liebig of Germany, called the "father" of soil chemistry, proved that treatment with strong acid increased the availability of bone nutrients to plants.

Necessity impelled the development of artificial fertilizers, which are chemical substances containing, in forms readily available to plants, the elements that improve the growth and productivity of crops. The three major nutrient elements that a crop needs are nitrogen, phosphorus, and potassium. (Other "minor" elements are required in much smaller amounts.)

The first artificial fertilizer was superphosphate, invented by the English agricultural chemist John Bennet Lawes. In 1842, he patented a process of treating phosphate rock with sulfuric acid to make the phosphate soluble, thus initiating the chemical fertilizer industry. Lawes later founded the world's first agricultural experiment station on his own estate of Rothamsted, not far from London. Potash fertilizer could be extracted in readily soluble forms, such as potassium chloride, from geological deposits found in several countries, including Germany, France, the USA, Canada, and from the brine of the Dead Sea.

The problem remained how to supply sufficient nitrogen to satisfy crop needs. Paradoxically, plants live in a veritable ocean of nitrogen—some 79% of the atmosphere—but are unable to assimilate it in its elemental form. During the latter part of the nineteenth and early part of the twentieth centuries, the major source of nitrogenous fertilizers were the saltpeter (sodium nitrate) deposits of Chile, and the guano (accumulated dung of seabirds) in the islands off the shore of Peru. The need to mine and transport these substances across the ocean, and the frequent disruption of international trade by wars in the twentieth century, made these fertilizer sources too expensive and insecure.

The long-range problem of supply was solved just before World War I by the work of Fritz Haber and Carl Bosch in Germany. Haber discovered a way to synthesize ammonia virtually from air and water, by getting elemental nitrogen to combine with hydrogen under high pressure and moderately high temperature. The process, as subsequently refined and industrialized by Bosch, requires energy that is generally obtained from fossil fuels.

The advent of chemical fertilizers marked a revolutionary change in modern agriculture. Along with improvement of crop varieties and of methods to control diseases and pests, as well as to prevent soil erosion, the development of fertilizers brought about a dramatic increase in crop yields. In a real sense, it fulfilled Jonathan Swift's observation quoted at the beginning of this chapter. On the other hand, it accelerated the spiral of population growth, enhanced by modern medicine and hygiene. Moreover, the excessive use of fertilizers eventually began to cause environmental pollution and degradation. Thus, we have yet another example of how an innovation designed to alleviate one problem—if applied injudiciously—may beget other problems[1].

---

[1] The subsequent careers of Fritz Haber and Carl Bosch both took tragic turns. In World War I, Haber helped the German war industry to manufacture explosives, as well as to develop and apply chlorine gas that was used to poison Allied soldiers. Haber's wife then committed suicide. After the war, the Allies declared him to be a war criminal. Nonetheless, he was awarded the Nobel Prize in 1919 for his invention of ammonia synthesis. All his patriotic zeal for the German Fatherland, ironically, availed him naught when, in 1935, he was hounded out of Germany with the epithet "the Jew Haber". He died in exile the same year. Carl Bosch, for his part, helped to develop a process to convert coal to synthetic gasoline, a fuel that eventually helped Hitler to wage World War II. Bosch received a Nobel Prize for chemistry in 1931, but when he opposed Nazi policies he was dismissed from his position at Farben Industries and died in 1940, before some of the chemical products of his company were used in the Holocaust.

12. SOIL BIODIVERSITY

God said: Let the earth bring forth
every kind of living creature...
and all kinds of creeping things of the soil...
and God saw that it was good.
Genesis 1:24–25

THE SUBSTRATE OF LIFE

The soil underfoot may seem inert, but is in fact full of life. It is not only the growth medium for higher plants (trees, grasses, and shrubs, including agricultural crops) but also home to a varied and interdependent biotic community consisting of myriads of organisms, including microscopic and macroscopic plants and animals, all engaged in a complex set of complementary functions.

Notwithstanding its typically variable conditions (alternating saturation and dryness, heat and cold, swelling and shrinking, acidity and alkalinity, increasing and decreasing concentration of solutes, et cetera), soils host the highest density and diversity of organisms of any domain in the biosphere. By some estimates, soils contain more than twice the number and variety of microbes that are present in the world's oceans, although the total volume of the oceans is hundreds of times greater than that of the world's soils. Though soil organisms are distributed throughout the active soil profile, their greatest density and diversity occurs near the surface, in the so-called "topsoil," generally having a depth not much greater than 20 cm (and often less than that).

A single cubic meter of topsoil may contain several billions of organisms, among them a billion fungi, ten billion or more bacteria and actinomycetes, ten million nematodes, thousands of earthworms, and millions of protozoa and algae. The soil may therefore be the biosphere's most vital and diverse domain. Practically every phylum known in biology is represented in the soil, each with a wealth of species diversity. That diversity is expressed not merely in the number of species or organisms present per unit volume, but indeed in the myriad interlinked functions performed therein. It is the very diversity of life that imparts to the soil the ability to respond to and recover from perturbations such as episodes of drought or flooding, as well as contamination by pollutants; that is to say, biodiversity enhances the soil's stability and resilience.

The soil community is engaged in an interdependent sequence of nutrient and energy transfers that constitute the soil's food web. Green plants are *primary producers*. Organisms feeding on living plant tissues (herbivores) are termed *secondary consumers*. Organisms that prey on them (parasites and predators) are called *tertiary consumers*. An additional set of organisms that feed on dead and decaying organic matter, known as *saprotrophic organisms*, function as *decomposers*. The distinctions among these functional groups, however, are not always mutually exclusive. Associations among different groups can be synergistic, where the species benefit from one another; or they can be antagonistic, where one species benefits at the expense of another. Through all these interconnections, soil biota influence the functioning of ecosystems.

The zone in the soil influenced by roots is called the rhizosphere. It extends a few millimeters from the surfaces of living roots. The rhizosphere develops along with the growing root system. After roots die, rhizosphere spaces often persist as a set of biopores, which may be utilized by subsequent root growth. Plant roots secrete various substances into the rhizosphere that enhance root-soil contact and in turn are utilized by microorganisms. Plant roots, by their own activity and by their association with microbial communities, modify soil conditions, such as the pH and the solubility (and hence availability) of nutrients.

Organic matter in the soil is both a product and a stimulator of biological activity. Many if not most of the organisms in the soil depend on the cycling of carbon compounds. The organic materials deposited in the soil by growing plants, as well as the processed remains deposited by animals that consume plants, are subsequently decomposed by microorganisms, which in turn release the carbon dioxide and nutrients that are once again assimilated by plants in the continuous interchange between the soil and plants that constitutes the dynamic functioning of terrestrial ecosystems.

SOIL FLORA

Higher plants, including crop plants, are termed *macroflora*. The root systems of such plants normally proliferate throughout the upper horizons of the soil profile. In favorable conditions, the length of roots permeating a cubic meter of topsoil may total hundreds of kilometers. Roots release acids and other substances that enhance nutrient availability and that spur the activity of microorganisms in the region immediately surrounding the roots and altered by their activity.

Microscopic plants such as algae and cyanobacteria are classed as *micro-flora*. They are photosynthetic organisms, mostly active at or near the soil surface. Some of these organisms can contribute nitrogen, as well as exude polysaccharides that serve to stabilize soil aggregates. Other microscopic forms include fungi, lichens, bacteria, and viruses.

Algae

Algae are autotrophs, able to perform photosynthesis, hence they require sunlight and tend to concentrate at or near the soil surface, to which they may impart a greenish hue. Their cells are of the order of 10 to 40 micrometers in diameter, and they occur either individually or in clusters (or colonies). Algae produce polysaccharides, substances that serve to bind and stabilize soil aggregates.

On exposed rock surfaces, algae may associate with fungi, forming symbiotic colonies known as *lichens* that also promote the gradual weathering of rocks. Such colonies are generally quite hardy, able to survive harsh conditions (including long periods of dryness, cold, or heat such as occur in deserts and in polar regions). However, lichens tend to be rather sensitive to air pollution (especially to the presence of sulfur dioxide, causing acid rain). Some lichen associations may also include cyanobacteria.

Fungi

The fungi in the soil include both microscopic and macroscopic yeasts, moulds, and mushrooms. Some of the fungi produce long chains of cells called *hyphae*, and these interweave to form branching filaments also known as *mycelia* that – though only a few micrometers in width – may extend many centimeters or even meters in length. The filamentous fungi permeate and decompose organic residues (including plant litter and animal remains). The fruiting bodies of some fungi appear as toadstools and mushrooms. Fungi are typically saprophytic (deriving their energy and nutrients from dead tissues), but some (e.g., Verticillium and Fusarium) infest living plant roots. Fungi serve to decompose organic matter and may exude various active substances that help to mobilize and transmit nutrients.

Some of the fungi enter into symbiotic association with the roots of higher plants by forming colonies that enmesh and even penetrate plant roots, thus forming fibrous or gelatinous coatings around the roots, called mycorrhizae, While extending and reaching into small pores and other regions of the soil that would otherwise be inaccessible to roots, they assist the plants to absorb nutrients from the soil. In this symbiotic relationship, these fungal colonies obtain vital substances (sugars, starches, proteins, and lipids) that are produced by the plants.

The mycorrhizae consist of two types. The first are *ectomycorrhizae*, which produce fruiting bodies as mushrooms that emerge above ground near the tree with which they are associated. A large colony of mushrooms may in fact constitute a single organism, with extensive underground linkages. The second type of mycorrhizae are *arbuscular mycorrhizae*, which remain

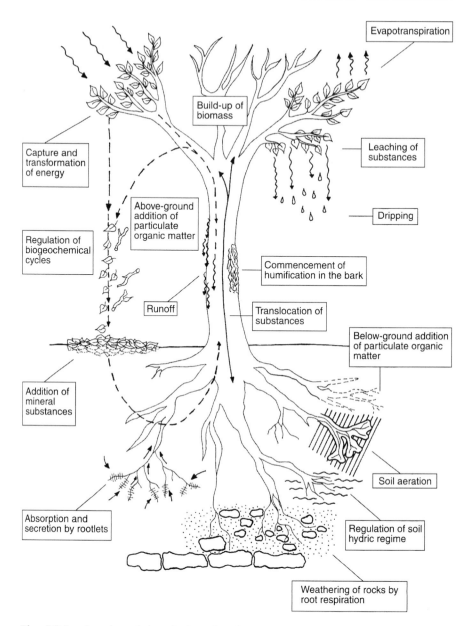

Fig. 12.1. Functions of plants in the soil (Gobat et al., 2004).

entirely underground. Their hyphae puncture the root tips of plants and are thin enough to penetrate narrow spaces in the soil that roots cannot enter. By forming vast networks, which in a cubic meter of soil may total thousands of kilometers in length, these strands can draw water and nutrients, including phosphorus and nitrogen, from soil regions that the roots themselves could not reach otherwise.

Bacteria

Bacteria are unicellular organisms that are either rod-shaped (called bacilli), spherical (cocci), or spiral-shaped. They range in size from 1 to several micrometers in length or width. They are not uniformly distributed throughout the soil but tend to proliferate and concentrate in regions where physicochemical conditions are favorable and their food is relatively abundant (e.g., near roots or concentrations of organic matter). Some bacteria are autotrophs. A class of bacteria known as *cyanobacteria*, formerly classified as *blue-green algae*, are a specialized group that live at or near the soil surface and are able to conduct photosynthesis. However, most of the bacteria in the soil are heterotrophs and live by the decomposition of plant and animal residues.

Bacteria are a highly diverse group of organisms, consisting of many thousands of species, and they tend to be abundant in the soil. Their total biomass varies from less than 200 to more than 1500 kg per hectare. They perform many functions within the soil. They decompose organic matter (either aerobically or anaerobically), recycle nutrients, and perform oxidation and reduction reactions.

A particularly important set of functions pertains to the cycling of nitrogen. Some bacteria are capable of absorbing elemental nitrogen from the atmosphere and combining it with hydrogen to form ammonia, which can then be absorbed as a nutrient by plants. Of these so-called nitrogen-fixing bacteria, some are free-living and some coinhabit the soil in symbiotic association with the roots of legumes (from which the bacteria, in turn, receive carbohydrates for energy).

Actinomycetes

Actinomycetes are microorganisms intermediate in form and function between bacteria and fungi. They are heterotrophic organisms that thrive

Table 12.1   Groups of Organisms Present in the Soil (Wild, 1994)

Microorganisms in fertile soil[a] (millions g⁻¹)		Animals in mull soil under beech	
		Millions ha⁻¹	% of total animal mass[b]
Bacteria	1–100	Earthworms 1.8	75.1
Actinomycetes	0.1–1	Enchytraeid worms 5.3	1.5
Fungi	0.1–1	Gastropods 1.0	7.0
Algae	0.01–0.1	Millipedes 1.8	10.6
Protozoa	0.01–0.1	Centipedes 0.8	1.8
		Mites and springtails 44.1	0.4
		Others 7.2	3.6

[a]The carbon in soil microorganisms accounts for about 3% of the soil organic carbon; that is, they contain about 1 tonne of carbon per hectare in a soil with 1.5% carbon to 15 cm depth.
[b]Total mass of animals was 286 kg.

on decomposable organic matter and proliferate in soils that are rich with plant and animal residues. They are most prevalent in warm, aerobic soils. Actinomycetes typically form branched filamentous networks and are capable of decomposing recalcitrant compounds such as cellulose. These organisms commonly release the earthy aroma often noticed when spreading compost or hoeing humus-rich soil.

A well-known genus of actinomycetes is *Streptomyces*, from whose products the medical antibiotics *streptomycin* and *aureomycin* were originally extracted in the 1940s.

SOIL FAUNA

Fauna that are less than 1 mm in size are known as *microfauna*. They are represented in the soil by numerous morphologically diverse groups. Prominent among them are single-celled organisms called protozoa, which include amoebae, ciliates, and flagellates. They are especially active in moist conditions and in the near vicinity of plant roots, and they tend to prey on bacteria, fungi, and other organisms. Their activity helps to release nutrients and to stimulate root growth.

Larger animals (albeit still small enough to be categorized as microfauna) are nematodes. These are microscopic worms that generally feed on bacteria and fungi. Among the nematodes are some that infest plant roots, thereby making them vulnerable to attack by other pathogens. Some crop plants (including corn, soybeans, and sugar beets) are especially vulnerable to nematodes, whereas others (e.g., marigolds and canola) produce root exudates that repel nematodes.

Macrofauna in the soil also include large invertebrates such as earthworms, termites, ants, millipedes, centipedes, spiders, beetles, and mites. A particularly

---

 BOX 12.1   Bacterial Nitrogen Fixation

Over a century ago, soil scientists in Germany and in Russia noticed that some leguminous plants (including peas, beans, lupines, clovers, and vetch) are able to thrive in nitrogen-deficient soils, where other crops could hardly grow. The legumes failed to grow, however, If the soil were sterilized. They concluded, correctly, that some organism in the soil supplies those legumes with nitrogen. In 1888, a class of bacteria was identified by the Dutch microbiologist Martinus Beijerinck. These bacteria, named *Rhizobium*, evidently attach themselves to the roots of leguminous plants, forming nodules. They are also able to absorb elemental nitrogen from the air and transform it into ammonia, which can then be made available to the host plants. Consequently, methods were developed to inoculate seeds of legumes with the appropriate bacteria in order to facilitate crop growth.

In the course of time, several other types of microorganisms have been identified as capable of nitrogen fixation, either symbiotically or independently. Notable among the nitrogen-fixing bacteria are the members of the family called *Cyanobacteria*, which are able to fix both carbon from the atmosphere by photosynthesis and nitrogen by conversion to ammonia.

prominent role is played by earthworms, which ingest huge quantities of soil as they tunnel and wend their way through the soil, and in so doing expel nutrient-rich casts while creating an extensive network of pores that improve aeration and infiltration. Thousands of species of earthworms have been defined in various regions, ranging in size from several millimeters to one meter. They are most common in humus-rich soils.

Termites abound in the soil mainly in tropical and subtropical regions. They dig labyrinthine tunnels in the soil and build huge chambered mounds above the ground's surface. Their burrows facilitate the internal drainage of water and the aeration of the soil. Similar works are constructed by ants, whose intricate and extensive colonies reach deep into the soil. Dung beetles, where they are active, help to distribute and decompose manure deposited by larger animals, such as grazing ungulates. Tunnels dug into the soil profile by such burrowing animals as gophers and often filled with soil from different horizons are called *crotovinas*. The formation of such features is an important mechanism by which the profile is mixed and the fertility of its deeper layers is enhanced.

Macrofauna living in the soil also include vertebrates (e.g., lizards, snakes, and various burrowing rodents such as gophers, field mice, moles, pack rats, and prairie dogs).

MAINTAINING SOIL BIODIVERSITY

Soil biodiversity is generally a characteristic quality of natural ecosystems. The conversion of virgin soil to cultivation typically causes a great loss of

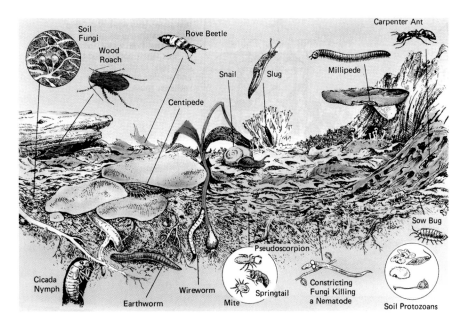

Fig. 12.2. Soil organisms (Nebel 1987).

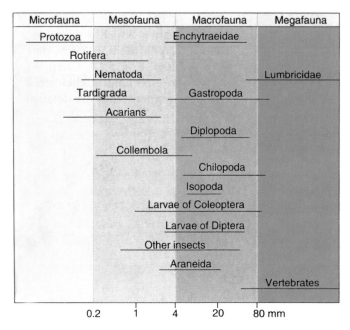

Fig. 12.3.   Microfauna, mesofauna, macrofauna, and megafauna in the soil (Gobat et al., 2004).

BOX 12.2   The Soil as an Ecosystem

An ecosystem is a community inhabiting a common domain in which living organisms function in mutual interaction with one another and altogether with their shared environment. A biotic community within a terrestrial ecosystem consists of *primary producers* (photosynthesizing green plants), *primary consumers* (herbivores), *secondary consumers* (carnivores), and *decomposers* (*detrivores*, generally microorganisms). The primary producers are able to synthesize their food by combining carbon dioxide obtained from the air with water and nutrients obtained from the soil, and using energy absorbed from sunlight. The products of photosynthesis are successively consumed by herbivores and then by carnivores, who utilize the energy contained in those products in the process of respiration. The closing stage in the cycle of energy and carbohydrate synthesis is carried out by decomposers, mostly by microorganisms that derive their energy by reoxidizing the carbon and hydrogen in carbohydrates while emitting carbon dioxide.

The crucible of this cycle on land is the soil. An important attribute of a healthy soil is its biodiversity, with a community of innumerable micro- and macroorganisms performing complementary tasks. The danger is that human intervention designed to control weeds, insects, or diseases may affect biodiversity in ways unforeseen, and also may contaminate groundwaters and surface waters.

BOX 12.3   The Role of Earthworms

Earthworms have aroused the interest of such diverse observers as Aristotle and Darwin. The role of earthworms in the soil was described most vividly and aptly two centuries ago by George White, as quoted by Russel (1912):

Worms seem to be the great promoters of vegetation, which would proceed but lamely without them, by boring, perforating, and loosening the soil, and rendering it pervious to rains and fibers of plants, by drawing straws and stalks of leaves and twigs into it, and, most of all, by throwing up infinite numbers of lumps of earth called wormcasts, which being their excrement, is a fine manure of grain and grass... The earth without worms would soon become cold, hardbound, and void of fermentation, and consequently sterile.

A population of several million earthworms (of various species, prominent among them being *Lumbricus terrestris*) per hectare is common where the supply of moisture and fresh organic matter is adequate. Such a population of earthworms can digest and expel as "casts" many tons of soil per hectare yearly.

Lumbricus terrestris is one of the most ubiquitous species of earthworms. It ingests organic debris (fallen leaves and dead roots) and may burrow into the soil to a depth of one or two meters to escape desiccation. It is also sensitive to aeration restriction, so that when the soil is drenched by a heavy downpour, this earthworm rises to the surface by the droves. It is relatively tolerant to changes in soil acidity but depends greatly on the abundance of organic matter (hence, it is most prevalent in grasslands), which it churns and mixes within the soil. In the process, it creates channels that serve to facilitate the infiltration of water and the penetration of air into the soil profile.

native biodiversity. That loss is direct, inasmuch as the mode of cultivation is aimed at promoting the production of a single crop, or, at most, at a specific sequence of crops, at the expense of all organisms that do not directly serve that aim. At the same time, the loss of biodiversity can accrue indirectly as an unintended consequence of soil and ecosystem degradation, including compaction, erosion, leaching of nutrients, burning of residues, unreplenished loss of organic matter, excessive fertilization, salination, contamination by pesticides and by sewage, acidification or alkalization, and the invasion of exotic species. The loss of the original species diversity pertains not only to higher plants, but indeed to soil microbes, mycorrhizae, nematodes, termites, and all other natural forms of microscopic and macroscopic life.

Although it is generally easier to maintain soil and ecosystem integrity than to restore it once it has been degraded, certain practices offer the promise of ameliorating disturbed soils. A notable and laudable trend that has gained currency in recent decades is the move toward minimum tillage. The traditional practice of "clean" plowing and repeated tillage by means of heavy tractors and soil-pulverizing implements, preformed annually over many decades, has resulted in much cumulative damage to soil structure and biodiversity. The reverse of that trend is the move toward "no-till" or "low-till" farming. This is a set of practices that minimizes disturbance of the soil and that maintains and augments organic matter over and in the soil. Consequently, energy is saved, soil structure is improved, infiltrability and aeration are enhanced, erosion

BOX 12.4   Useful Products of Soil Organisms

Realizing the rich diversity of life forms and products in the soil, scientists have searched that veritable treasure-trove for various beneficial uses. The study of soil microbiology in the early 1940s led to the discovery that some soil organisms produce substances that can serve medicinal purposes. The term *antibiotics* was coined in the early 1940s by microbiologist Selman Waksman, who, together with his students at Rutgers University, extracted actinomycin and streptomycin from actinomycetes found in the soil. Subsequently, the search has been on for soil organisms and products to help remediate pollution by petroleum products and other harmful substances.

Another application of microorganisms involves the extraction of a toxin from the bacterium *Bacillus thuringiensis* (Bt). This naturally produced toxin has been incorporated into pesticide formulations that can be sprayed on crops to control herbivorous insects, thus obviating the need for applying synthetic pesticides that are persistent in the environment and can cause unintended damage to the ecosystem. More recently, the gene that produces the Bt toxin has been introduced by bioengineering methods into certain crops in order to make the plants themselves repellent to insects. Such genetic manipulation, however, is not without attendant dangers and must therefore be regulated with utmost care.

is minimized, and carbon is sequestered in the soil. The problem, however, is how to control weed growth without having to rely on excessive use of herbicides.

BIOLOGICAL NITROGEN FIXATION AND CYCLING

Biological nitrogen fixation is an extremely important process in the soil, carried out by several groups of bacteria and actinomycetes able to absorb elemental nitrogen from the atmosphere and combine it into compounds that serve as nutrients from plants. These microbial organisms produce the enzyme nitrogenase that catalyzes the reduction of dinitrogen gas (N_2) to ammonia. That process requires energy. The nitrogen-fixing microorganisms are either free-living or symbiotic.

Some of the free-living nitrogen-fixing organisms obtain energy directly from sunlight, and others derive it from soil organic matter. Prominent among the latter is a genus named *Azotobacter*, which is an aerobic heterotroph. Other organisms (e.g., members of the genus *Clostridium*) are active in anaerobic conditions. Although both groups of free-living organisms fix only small amounts of nitrogen (not much more than 1 kilogram per hectare annually), they can be important in sustaining plant communities in natural ecosystems.

A different group of nitrogen-fixing microorganisms is symbiotic. These organisms obtain their energy from their host plants, to whose roots the microorganisms attach themselves, forming nodules. Important members of this group are the bacteria classified as *Rhizobium* and related genera. They enter into symbiosis with various species of the agriculturally important

Leguminosae family. These include herbage plants of the genera clover (Trifolium) and alfalfa (Medicago); as well as the food crops soybeans (Glycine max), peanuts (Arachis hypogaea), peas (Pisum), beans (Phaseolus), and various others. Properly managed, leguminous crops—thanks to their symbiotic nitrogen-fixing capacity—can help to maintain and even enhance soil fertility.

Other organisms, such as the actinomycete genus *Frankia*, are known to associate similarly with several other families of plants. The latter even include species of tree genera such as the temperate-region *Alnus* and *Myrica*, the arid-region *Acacia*, and the tropical-region *Casuarina* and *Ceanothus*. In the latter region, efforts are being made to develop a crop-rotation system with agro-forestry that utilizes leguminous trees (e.g., *Leucaena leucocephala*) able to incorporate significant amounts of nitrogen into the soil for the subsequent benefit of crop production.

Cyanobacteria, formerly classified as blue-green algae, are able to fix nitrogen both in the free-living mode and in symbiosis with higher plants. As such, they may have been among the first colonizers of land, some two or more billion years ago. They are active especially in shallow aquatic media, such as marshes and flooded rice fields. In the latter medium, an organism named *Anabaena* lives in association with a water fern named Azola, and their symbiosis can produce as much as 50 kilograms of nitrogen per hectare annually, enough to fertilize a productive crop of rice.

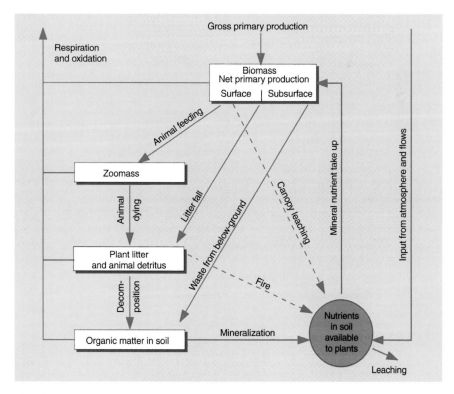

Fig. 12.4. Functions of soil organisms (Schultz, 2002).

Processes that are important stages in the cycle of nitrogen in soils are *nitrogen fixation*, *ammonification*, *nitrification*, and *denitrification*. *Nitrogen fixation* is the absorption of elemental nitrogen from the atmosphere (N_2) and its incorporation in organic compounds of living organisms. *Ammonification* is the formation of ammonium (NH_3 and the ion NH_4^+) from nitrogen-containing organic compounds such as amines ($-NH_2$). *Nitrification* is the oxidation of ammonium to nitrite (NO_2^-) and nitrate (NO_3^-). *Denitrification* is the reduction of nitrogen oxides back to the elemental form. These processes can be regarded as stages in a single continuous reciprocal process, namely the *nitrogen cycle* in soils. Each stage of the overall process is carried out by a specialized group of microorganisms, and is greatly influenced by the condition of the soil; i.e., whether it is dry, moist, or saturated, aerobic or anaerobic, cold or warm, with an abundance or a paucity of organic matter, etc.

The cycle of nitrogen in the soil and its exchange between the soil and the atmosphere can have a bearing on atmospheric processes as well. For example, partial denitrification in the soil releases nitrous oxide, which is a powerful greenhouse gas implicated in global warming. Other greenhouse gases that are released from the soil in certain circumstances are carbon dioxide and methane.

13. SOIL AND WATER MANAGEMENT

By the sweat of your brow
shall you eat bread
Genesis 3:19

MODES OF CULTIVATION

Proper soil management in agriculture consists of a series of practices that include cultivation, planting, fertilization, pest control, irrigation, drainage, and erosion control, The more efficiently these practices are carried out and optimized, the more productive and sustainable will agriculture become.

Cultivation or *tillage* is usually defined as the mechanical manipulation of the soil aimed at improving conditions affecting crop production. Three principal aims are generally attributed to tillage: control of weeds; incorporation of organic matter into the soil; and improvement of soil structure. An additional rationale is sometimes claimed for tillage, namely the conservation of soil moisture by enhancing infiltration and inhibiting evaporation.

A distinction must be made between primary tillage and secondary tillage. *Primary tillage* is typically carried out by means of moldboard plows or disk plows, both of which slice and lift the soil along parallel furrows and invert it so as to cover the surface residues. Subsoilers and chisels, also used for primary tillage, break and loosen the soil without inverting it. All such methods of primary tillage are generally designed to penetrate to a depth of at least 20 cm, and sometimes to a depth as great as 50 cm.

Secondary tillage is carried out in some cases subsequent to primary tillage, to repeatedly loosen the soil and eradicate weeds. In other cases, secondary "light" tillage is performed in lieu of primary tillage in soils that are naturally

175

loose and require no primary tillage at all. As such, secondary tillage aims to loosen the soil to a relatively shallow depth, generally less than 20 cm. The implements suitable for secondary tillage are disk harrows, spike harrows, seeps, rotary hoes, cultipackers, and various other tools that work the soil to shallow depth and help to disrupt crusts where they occur. All too often, however, such implements are efficacious in the short run (e.g., in preparing a seedbed) but ultimately contribute to the degradation of soil structure by grinding down the soil's natural aggregates.

In recent decades, the advent of chemical herbicides has reduced the importance of tillage as the primary method for the eradication of weeds, though the high cost of such chemical treatments and their ancillary environmental effects limit their application, especially in developing countries. At the same time, the formerly prevalent practice of inverting the topsoil in order to bury manures and plant residues has become a less important function of tillage in modern field management. Plant residues can, and in many cases should, be left over the surface as a *stubble mulch* to protect against evaporation and erosion.

An essential task of agriculture is *soil structure management*, as it affects water infiltration and runoff, wind erosion and evaporation, gas exchange processes, as well as planting and germination of crops. Here we find that tillage practices suitable in one location may become harmful in another. Arid-zone soils with low organic matter contents and unstable aggregates are particularly vulnerable to compaction, crusting, and erosion. The precise effects of various modes of tillage must be defined in each case for tillage to be practiced efficiently and sustainably.

Tillage operations are especially consumptive of energy. The amount of earth-work involved in repeatedly loosening, pulverizing, inverting, and then recompacting the topsoil is indeed very considerable. In a typical small field of 1 hectare, the topsoil to a depth of only 30 cm weighs no less than 4000 tons.

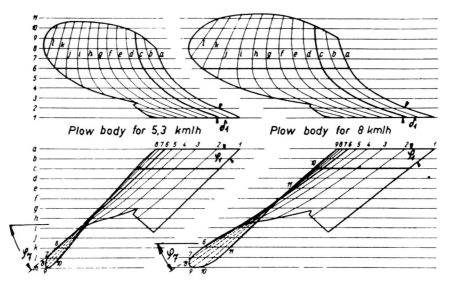

Fig. 13.1. Universal shape of a moldboard plow for deep primary tillage.

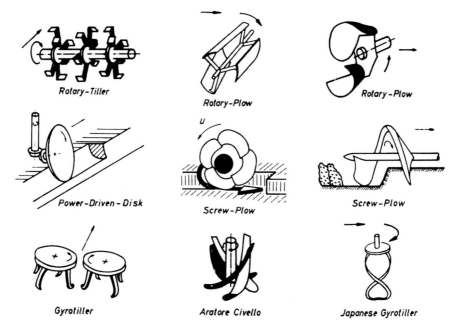

Fig. 13.2. Horizontal and vertical rotating harrows for shallow secondary tillage.

In an extensive farm of 1000 hectares, the mass of soil thus manipulated in each cycle of operation may exceed 4 million tons. The consumption of energy, as well as the wear and tear of tractors and implements, increases steeply as the depth of tillage increases. With the rising cost of fuel, the costs of tillage also increase progressively. Moreover, much damage is done to soil structure by the repeated passage over the soil of heavy tractors and other machinery, and such damage, which affects infiltration, aeration, germination, and root system development, is difficult to rectify.

Recent trends in soil management are aimed at minimizing tillage operations and travel, both to reduce costs and to avoid soil compaction, while tailoring each operation to its specific zone and objective. This approach, in numerous variations, underlies the methods variously termed "minimum tillage", "precision tillage", and even "zero tillage" ("no-till" in common parlance). However, methods developed in one location may not be suitable for another location, where soil and climate conditions and economic constraints differ greatly. Some, but not all soils, have favorable structure (called "tilth" in the classical farming terminology) quite naturally and require very little is any tillage. Others, however, develop hardpans such that inhibit root proliferation and hence can be improved by appropriate tillage.

An important current trend is to adopt a comprehensive system of soil and crop management called "precision farming". It consists of a balanced combination of practices designed to optimize nutrient supply, tillage, water use, and pest control. Instead of treating a large unit of land uniformly, it recognizes each field's inherent heterogeneity. Accordingly, it relies on remote sensing and monitoring of the field to determine the space-variable and time-variable requirements for all inputs and interventions. Tractors and ancillary machines traversing the field are provided with precise data regarding the spot-to-spot needs for applying pesticides, fertilizers, seeds, and water, and with automated means for responding to those needs continuously.

A related set of practices designed to maintain and even enhance soil productivity while minimizing energy consumption is called "minimum tillage" or even "zero tillage". The idea is to avoid the traditional practice of "clean cultivation" of the entire top layer of the soil, which consists of burning or plowing-in the stubble of previous crops and disrupting the natural structure of the soil, thus making it more vulnerable to erosion. Instead, special equipment is used that is designed to sow seeds into narrow slits while retaining the residues on the surface. Those organic remains, called "mulch", help to conserve moisture and protect the soil against both wind erosion and water erosion. The problematic aspect of zero tillage is that it relies on the use of herbicides instead of mechanical cultivation to control the weeds that might otherwise compete with crop plants for moisture, nutrients, space, and light.

 BOX 13.1   Shifting Cultivation

Shifting cultivation is a mode of farming long followed in the humid tropics of Sub-Saharan Africa, Southeast Asia, and South America. In the practice of "slash and burn", farmers would cut the native vegetation and burn it, then plant crops in the exposed, ash-fertilized soil for two or three seasons in succession. As the original organic matter reserve in the topsoil decomposed and as the high rainfall would leach out the nutrients from the root zone, the farmers would abandon the cleared plot and move to an adjacent patch of forest. They would allow each cultivated plot to recover its vegetation and fertility for some fifteen or twenty years before returning to it. Thus, they practiced an extensive rotation (forest-crop-forest) that was sustainable for many generations, while the population density remained low. What disrupted the system was the progressive growth of population that has taken place in the last century. Population pressure has forced farmers to return to the same plots earlier before the soil had been given the time to be completely rejuvenated. Soil fertility then began to deteriorate, owing to the extraction of nutrients without replenishment and to progressive erosion of the bared soil.

SOIL COMPACTION

In the agronomic sense, a soil is considered to be compacted when the total porosity (in particular, the air-filled porosity) is so low as to restrict aeration, as well as when the soil is so tight, and its pores so small, as to impede root penetration, infiltration, and drainage. Soils may become compacted naturally as a result of the way they were formed in place. Surface crusts may form over-exposed soils under the beating and dispersing action of raindrops, followed by drying and hardening of that surface layer. Naturally, compact subsurface layers may consist of densely packed granular sediments, which may be partially cemented. Indurated layers, called *hardpans*, can be of variable texture and, in extreme cases, may exhibit rocklike properties. Such indurated layers, called *fragipans*, may become almost totally impenetrable to roots, water, and air. They form typically at the junction of two distinct layers of soil, where penetration of water and/or dissolved or suspended materials is retarded by a clay layer, a water table, or bedrock.

A *claypan* is a tight, restrictive subsoil layer of high clay content that tends to be plastic and relatively impermeable to water and air. In humid climates, such layers may remain perpetually wet and give rise to perched water tables above them, thus inducing anaerobic conditions within the root zone. In the case of alluvial soils (originally deposited by water, generally in river valleys), claypans may be depositional layers; in other cases, they may have developed *in situ* as a result of clay translocation from top layers and accumulation at some depth within the profile.

Quite apart from the natural formation of compacted layers, the occurrence of soil compaction in agricultural fields can be due to the influence of mechanical forces applied at or near the soil surface. One such cause of soil compaction is trampling by livestock. However, by far the most common cause of soil compaction in modern agriculture is the use of heavy machinery, including tractors and other vehicles, as well as soil-engaging implements. The magnitudes of pressures exerted on the soil by wheeled and tracked vehicles depend in a combined way on characteristics of the soil and of the wheels or tracks involved. Lugs or ribs on tires or chains have the effect of concentrating high pressure, as well as of kneading the compressed volume of soil, thus causing severe damage to the soil's natural structure.

Pressures are imparted to the soil not only by vehicles traveling on its surface, but also by tillage tools operating beneath the surface. As tools of various designs are thrust into and through the soil, several different effects may occur simultaneously as the soil is cut, compressed, sheared, lifted, displaced, and mixed. Some of the soil is pushed ahead of the moving tool against the resistance of the static soil body and is thus also compacted. Consequently, the pressures acting on any volume element of soil in the field are seldom, if ever, isotropic (equal in all directions). Any differences between the principal stresses (i.e., the stresses acting along the horizontal, lateral, and vertical axes) necessarily give rise to shearing, as well as compressive, stresses. The simultaneous application of both compression and shearing contributes significantly to soil compaction. Yet another factor that comes into play during compaction is the change in matric potential of soil moisture resulting from the change in porosity and pore-size distribution. Soil compaction tends to raise the potential energy of soil moisture (i.e., to reduce the matric suction), a change that absorbs some of the mechanical energy imparted to the soil by compaction.

The effects described have an important bearing on the design and operation of off-road (cross-terrain) vehicles, such as recreational, exploration, and military vehicles. The engineering topic of *trafficability* pertains to the ability of variously shaped vehicles to traverse areas with disparate types of soil (sandy, clayey, gravelly, etc.) at a range of moisture contents in different topographies (from flat to steep).

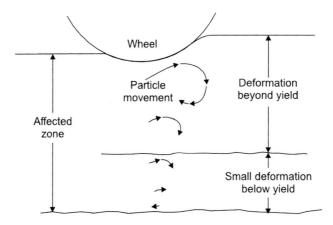

Fig. 13.3.   Soil deformation and displacement under a moving wheel.

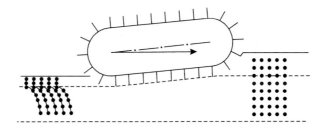

Fig. 13.4.   Compaction and shearing deformation caused by a moving tractor.

Control of Soil Compaction

A major task of soil management is to minimize soil compaction to the furthest extent possible. A secondary task is to alleviate or remedy the unavoidable measure of compaction caused by traffic and tillage after it occurs.

The most obvious approach to the prevention of soil compaction is the avoidance of all but truly essential pressure-inducing operations. This calls for minimizing tillage, and choosing the most efficient implements and the most appropriate time to apply them. Growing awareness of these requirements has led in recent decades to the development of integrated systems of conservation tillage. These systems also emphasize the importance of retaining crop residue as an organic matter-enriching mulch that not only improves soil structure but also protects the soil surface against rapid evaporation as well as against erosion by water and wind. In addition, reducing the intensity of tillage helps to conserves energy. However, excessive reliance on phytotoxic chemicals for weed control poses environmental problems owing to their possible persistence and the danger of contamination.

Since random traffic over a field by heavy machinery is a major cause of compaction, cultural systems have been developed to restrict vehicular traffic to permanent, narrow lanes and to reduce the fractional area trampled by wheels to less than 10% of the land surface. In row crops, seedbed preparation and planting can be confined to the narrow strips where planting takes place rather than be carried out over the entire surface as was the practice in former times.

An extremely important factor is the timing of field operations in relation to the state of soil moisture. Operations that impose high pressures should, if possible, be carried out on relatively dry soil, which is much less compactable than is moist soil.

BOX 13.2 Soil Structure Management

Good soil management is a set of practices that conserve and sustain soil productivity. That includes maintaining or improving the soil's physical condition (i.e., soil structure, aeration, water intake and retention) and its chemical attributes (near-neutrality of pH, supply and adequate concentration of nutrients, and absence of toxic factors).

Conventional methods of tillage generally consist of pulverizing and baring the entire soil surface each season so as to incorporate residues and kill weeds in preparation of planting a new crop. The advantage, in the short run, is the convenience of planting and the eradication of competing weeds. In the long run, however, the disadvantages can greatly outweigh the advantages. As the soil is bared and loosened, and as it is trampled repeatedly, it is exposed to scouring by rain and deflation by wind, and it tends to form a crust and a compact layer (a "plow-pan") that inhibits germination and root development. Erosion involves the loss of the most fertile layer of soil, called the "topsoil", which leaves the less fertile subsoil.

The alternative to conventional surface-clearing tillage is to maintain the surface-protecting residues ("stubble") of previous crops and to plant a new row-crop by opening narrow slits in the soil with minimal disturbance of the inter-row strips. This method is known as "no-till" or "minimum tillage" farming. It often requires, however, alternative methods of weed control, such as the application of herbicides by spraying. Such methods must be applied very carefully to the target area so as to minimize damage to plants in adjacent areas.

THE PRACTICE OF IRRIGATION

Irrigation is the supply of water to agricultural crops by artificial means, designed to permit farming in arid regions and to offset periodic droughts in semi-arid regions. Even in regions where total seasonal rainfall is adequate on average, it may be poorly distributed during the growing season and variable from season to season. Wherever traditional rain-fed farming is a high-risk enterprise, irrigation can help to ensure stable production.

Irrigation has long played a key role in feeding expanding populations and is expected to play a still greater role in the future. Although irrigated land amounts to only some 16 percent of the world's cropland, it contributes well over 30 percent of the value of agricultural production. That vital contribution is even greater in arid regions, where the supply of water by rainfall is least even as the demand for water, imposed by the bright sun and the dry wind, is greatest.

Irrigation can do more than merely raise the yields of specific crops; it can also prolong the effective crop-growing period in areas with extended dry seasons, thus permitting multiple crop growing per year where only a single crop could be grown otherwise. With the security provided by irrigation, additional inputs needed to intensify production further (e.g., pesticides, fertilizers, improved varieties, physiological treatments, environmental controls, soil amendments, and tillage) may become economically feasible. Irrigation reduces the risk of such expensive inputs being wasted by crop failure resulting from lack of water.

The practice of irrigation consists of applying water to the part of the soil profile that serves as the root zone, for the immediate and subsequent use of the crop. Inevitably, however, the initiation and the continuation of irrigation in a given area induce a series of processes that can profoundly affect the both the on-site and the related off-site environments, and not necessarily for the better. Over time, some of its potentially self-destructive effects may make the very practice of irrigation unsustainable.

Off-Site Aspects of Irrigation

The first requirement of an irrigation project is a dependable supply of fresh water and the means for its delivery to the site to be irrigated. The second requirement is the availability of suitable land and soil in which to grow the crops under irrigation. The third requirement is an outlet for the safe disposal of wastewater from the irrigated land. Every irrigation project therefore consists of withdrawing water from some source (a lake, an aquifer, or a river) and diverting it to the site of irrigation. In the process, inevitably, that water is denied to some other site that had been its natural recipient. In the typical case of a river valley project, changes take place both in the upstream and in the downstream sections of the riverine domain.

River flow is generally time-variable. More often than not, the season of peak irrigation demand coincides with the period of low river flow. Hence, an irrigation project typically requires the construction of engineering structures—dams and canals—designed to regulate the flow so as to ensure adequate storage and supply throughout the growing season.

Dam construction is a problem in itself. Appropriate locations for dams are relatively rare. The ideal topographic, geologic, and climatic conditions are seldom found in the proximity of target irrigation projects, and less-than-ideal conditions may make the construction and maintenance of dams economically prohibitive and environmentally unsustainable. Unfavorable topography may require very massive construction and may result in the submergence of very large areas with consequent damage to natural ecosystems, displacement of long-established human populations and infrastructure, and loss of historical cultural heritage and scenic sites. Some areas are inherently vulnerable to potential natural disasters such as earthquakes. Porous or fractured substrata may cause great losses of water by practically uncontrollable seepage, and a dry climate may impose additional losses of water by evaporation. A case in point is the Aswan Dam and its Lake Nasser, located in the midst of the earth's driest desert, where evaporative losses of water may be in the range of 12–16 billion cubic meters per year—some 20 percent of the inflow.

Additional losses of water by evaporation, and especially by seepage, occur in the system of conveyance from a dam to the fields. Generally, this conveyance takes place via canals and ditches, which, in the interest of minimizing costs, are often dug into the ground and left unlined; i.e., not underlain with concrete or some other impervious and scouring-resistant material. Even where closed conduits (i.e., metallic, ceramic, or concrete pipes) are installed, they tend in time to develop leaks and to incur considerable losses of water and hydraulic pressure.

Water storage behind dams is also subject to silting, which, gradually over a period of some decades, reduces a dam's capacity and may eventually clog up its storage basin entirely. Silt accumulation in reservoirs and canals is especially rapid in regions where the upper watersheds of the rivers have been denuded of their natural vegetative cover by overgrazing and further destabilized by excessive cultivation, thus subjecting the catchments to accelerated soil erosion.

Thus far, we have considered environmental processes that may take place upstream of an irrigated area. In addition, irrigation also entails a series of processes that occur downstream. The first of these is the diminution of the river flow resulting from the extraction of the water used for irrigation. Consequently, all the associated riparian ecosystems along the riverbanks, floodplain, and estuary are deprived of vital water supplies and thereby impoverished. Natural wetlands that were originally biologically diverse and highly productive may be subject to periodic or permanent desiccation, and even to the eradication of vital species or biotic habitats, especially fisheries. Where a river discharges naturally into a freshwater lake, the deprivation of the lake's inflow resulting from the river's diversion for irrigation may cause the lake to shrink markedly, and in places drastically. The shrinkage of the Aral Sea in central Asia is a prime example of the sort of environmental disaster that can be caused by large scale irrigation development. The shrinkage of the Dead Sea (shared by Israel, Palestine, and Jordan) is another example.

These environmentally damaging processes are greatly exacerbated by the downstream disposal of irrigation-generated wastes. Irrigated lands typically generate drainage waters, which tend to be laden with salts as well as with residues of the fertilizers and pesticides that are often applied in excess to the irrigated crops. If the drainage from the bottom of the root zone percolates downward toward an aquifer, it may gradually contaminate its groundwater and make the aquifer unusable for humans. In extreme cases it may even become unsuitable for irrigation. Nitrates as well as chlorides may accumulate in groundwater underlying irrigated lands, to the extent of posing a health hazard to communities relying on wells. Other agents contained in the drainage from irrigated agriculture, as well as from households and industries, are various toxic and carcinogenic elements and compounds from pesticides. One example is the element boron, commonly present in detergents, which, even in relatively small concentrations, can be toxic to certain crops. Yet another example is the element selenium.

Where the drainage from irrigated lands is channeled through drainage ditches or pipes and discharged into the river downstream, it pollutes the water there and may make it unusable for people as well as harmful to natural fauna and flora. The alternative to discharging the drainage into the river is to convey it to the sea (which may be quite distant), to lagoons or wetlands (whose fauna and flora, however, may also be vulnerable to the pollutants), to environmentally isolated evaporation basins in the desert, or to very deep aquifers where the pollutants are diluted. All of these alternatives may be quite expensive to carry out, and perhaps unsustainable in the long run. The Kesterson Reservoir in California is an example where the discharge of drainage from irrigated areas into wetlands has resulted in damage to wildlife due to the accumulation of waterborne toxic elements.

Onsite Aspects of Irrigation

The twin processes of degradation that typically affect irrigated lands in river valleys and low-lying coastal plains, especially those in arid regions, are waterlogging and salination. Waterlogging results from the tendency of irrigators to apply a volume of water to the soil in excess of the amount of soil-water taken up by the crop. In part, this is a matter of necessity, to prevent the root zone from accumulating salts.

Irrigation water is never entirely pure; hence, the application of irrigation necessarily adds water-borne salts to the soil. Moreover, many arid-zone soils and subsoils contain natural reserves of salts, which are also mobilized by irrigation. Since crop roots typically exclude most salts, the salts left in the root zone tend to accumulate to the detriment of the crop, unless leached from the soil and driven downward by drainage.

Maintaining the balance of water and of salts in the root zone, however, is a delicate dynamic process. Water flowing downward below the root zone eventually reaches the water-table and augments the groundwater saturating the substrata. If the water table remains deep and the groundwater beneath it has its natural outflow, the balance of water and salts in the root zone can remain favorable to crop growth. If, however, the water table is shallow and the rate of natural groundwater drainage is slow, the addition of irrigation water from above will cause the water-table to rise. Sooner or later (perhaps within a few decades) the water-table approaches the soil surface. The root zone then becomes saturated with water and deprived of oxygen. Since most crop plants require oxygen in the soil for their roots to respire, the saturation of the root zone (a condition called waterlogging) itself restricts crop growth and, in the case of sensitive plants, may cause total crop failure.

The process of waterlogging is further exacerbated by soil salination. When the water table comes within about 1 meter or less of the surface, the process of soil salination is accelerated. Instead of percolating downwards and leaching away the salts, the salt-laden water—especially where the groundwater is naturally brackish, as is commonly the case in arid regions—percolates upwards to the soil surface, where it evaporates, leaving the salts behind. So much salt can accumulate at and near the surface as to render the soil sterile. In this manner, well-intentioned irrigation projects can, quite inadvertently, induce the salination of originally productive soil.

A further scourge of irrigation is the phenomenon of soil degradation due to *sodicity* (also called *alkalinity*), a condition caused by the specific effect of sodium ions adsorbed onto the electrostatically charged clay particles. As explained in a preceding chapter, colloidal clay particles generally exhibit a negative charge. When surrounded by an aqueous solution of electrolytic salts, such particles attract cations while repelling anions.

The adsorbed cations form a swarm surrounding each clay particle. These cations are exchangeable, in the sense that they can be replaced by other cations whenever the composition of the ambient solution changes. Divalent cations such as calcium and magnesium tend to compress the swarm of cations, thus allowing the particles to approach one another sufficiently to clump together and to form flocs, a process called *flocculation*, which contributes to the formation of a desirable soil structure. In contrast, monovalent cations such as sodium tend to diffuse farther away from the particle surfaces and thus to thicken the hydration envelope surrounding each particle. This, in turn, causes swelling and dispersion of the soil flocs. The latter process, called *deflocculation*, destroys soil aggregates and restricts the soil's permeability to water and air. When wet, a sodic soil becomes a slick and sticky mud; however, when dry, it hardens to form a tough crust with a typical pattern of cracks. This condition not only reduces the entry of water and air into the soil but also forms a barrier to the emergence of germinating seedlings and to the penetration of their roots.

DRAINAGE REQUIREMENTS

The term *drainage* can be used in a general sense to denote outflow of water from soil. More specifically, it can serve to describe the artificial removal of excess water, or the set of management practices designed to prevent the occurrence of excess water. The removal of free water tending to accumulate over the soil surface by appropriately shaping the land is termed *surface drainage* and is outside the scope of our present discussion. The removal of excess water from within the soil, generally by lowering the water table or by preventing its rise, is termed *groundwater drainage*, which is an integral aspect of sustainable irrigation management.

The artificial drainage of groundwater is generally carried out by means of *drains*, which may be ditches, pipes, or "mole channels", into which groundwater flows as a result of the hydraulic pressure gradients existing in the soil. The drains themselves are made to direct the excess water, by gravity or by pumping, to the *drainage outlet*, which may be a stream, a lake, an evaporation pond, or the sea. In some places, drainage water may be recycled, or reused, for agricultural, industrial, or even residential purposes.

Because drainage water may contain potentially harmful concentrations of salts, fertilizer nutrients, pesticide residues, and various other potentially toxic chemicals as well as biological pathogens, it is not enough to "get rid" of it. A major concern is the eventual consequence of its disposal. Therefore, the first requirement of drainage management is to provide a safe outlet for the effluent. The recent emphasis on modes of agricultural management that minimize chemical inputs may help to lessen the problem posed by the persistence of some of these chemicals in the environment.

Various theoretical and empirical methods have been proposed for designing the optimal drainage system for different sets of conditions, considering the attributes

Table 13.1   Prevalent Depths and Spacings of Drainage Tubes in Various Soil Types

Soil type	Hydraulic conductivity (m/day)	Spacing of drains (m)	Depth of drains (m)
Clay	1.5	10–20	1–1.5
Clay loam	1.5–5	15–25	1–1.5
Loam	5–20	20–35	1–1.5
Fine, sandy loam	20–65	30–40	1–1.5
Sandy loam	65–125	30–70	1–2
Peat	125–250	30–100	1–2

of the soil, the climatic and hydrological regime, and the crops to be grown. The ranges of depth and spacing generally used for the placement of drains in field practice are listed in Table 13.1. In Holland, the country with the most experience in drainage, common criteria for drainage are to provide for the removal of about 7 millimeters of water per day, and to prevent a water table rise above 0.5 meter from the soil surface. In more arid regions, because of the greater evaporation rate and groundwater salinity, the water table must generally be kept much deeper. In the Imperial Valley of California, for instance, the drain depth ranges from about 1.5 to 3 meters, and the desired water table depth midway between drains should be at least 1.2 meter. For fine-textured (less readily permeable) soils, the depth should be greater still, especially where the salinity risk is high. Since there is a practical and economic limit to how deep the drains can be placed, it is the density of drain spacing that must be increased in such circumstances.

THE SUSTAINABILITY OF IRRIGATION

The sustainability of irrigation is never to be taken for granted. Waterlogging and salination of soils, along with other degradation processes, not only caused the collapse of irrigation-based societies in the past, but are indeed threatening the viability of irrigation at present. The problem is global in scope. Decimation of natural ecosystems, deterioration of soil productivity, depletion and pollution of water resources, and conflicts among sectors and states over dwindling supplies and rising demands have become international problems closely linked with irrigation development.

Irrigated agriculture can be sustained only if and where certain stringent requirements are met. The requirements are effective prevention of upstream, on-site, and downstream environmental damage.

Although there will be cases where the costs of continuing irrigation may be prohibitive, especially if severe damage has already occurred, in most instances, the cost should be worth bearing. Investing in the maintenance of irrigation and the integrity of the environment in general can result in improved economic and social well being.

Irrigated agriculture must strive for a balance between the immediate need to maximize production and the ultimate need to ensure continued productivity in the future. It must also strive to achieve a harmonious interaction with the external environment, which includes both natural ecosystems and other human enterprises. More specifically, irrigation projects should ensure that water supplies of adequate quality are and will continue to be available, the salt balance and hence the productivity of the land can be maintained, the drainage effluent can be disposed of safely, public health can be safeguarded, and the economic returns can justify the costs.

The *sine qua non* of ensuring the sustainability of irrigation is the timely installation and continuous operation of a drainage system to prevent water-logging and to dispose safely of excess salts. All too often drainage creates off-site problems beyond the on-site costs of installation and maintenance, since the discharge of briny effluent can degrade the quality of water along its downstream course. Where access to the open sea is feasible, solving the problem is likely to be easier than in closed basins or in areas far from the sea. In those cases, the disposal terminus eventually becomes unfit for human use, as well as for wildlife. Hence the importance of reducing the volume and salinity of effluents by such means as improving the efficiency of water use, a task that in itself can bring economic and environmental rewards. Modern irrigation technology offers the opportunity to conserve water through reduced transport and application losses, coupled with increased yields per unit volume of water.

 BOX 13.3   Silt and Salt in Ancient Mesopotamia

Ancient Mesopotamia owed its prominence to its agricultural productivity. The soils of this alluvial valley are deep and fertile, the topography is level, the climate is warm, and water is provided by the twin rivers, Euphrates and Tigris. However, the diversion of river water onto the valley lands led to a series of interrelated problems.

The first problem was sedimentation. Early in history, the upland watersheds were deforested and overgrazed. The resulting erosion was conveyed by the rivers as suspended silt, which settled along the bottoms and sides of the rivers, thus raising their beds and banks above the adjacent plain. During periods of floods the rivers overflowed their banks, inundated large tracts of land, and tended to change course abruptly. The silt also settled in channels and clogged up the irrigation works.

The second and more severe problem was salt. Seepage from the rivers, the irrigation channels, and the flood-irrigated fields caused the water table to rise throughout southern Mesopotamia. Because all irrigation waters contain some salts, and because crop roots normally exclude salts while extracting soil moisture, the salts tended to accumulate in the soil and groundwater. And as the undrained water table rose it brought the salts back into the soil. The farmers of ancient Mesopotamia attempted to cope with the process of salination by periodically fallowing their land, and by replacing the salt-sensitive wheat with relatively salt-tolerant barley. However, the process proceeded inexorably. So the ancient hydraulic civilizations of Sumer, Akkad, and Babylonia each in turn, rose and then declined, as the center of population and culture shifted over the centuries from the lower to the central to the upper parts of the Tigris-Euphrates valley (Hillel, 1994).

BOX 13.4   How Ancient Egypt Escaped the Scourge of Salinity

In contrast to Mesopotamia, the civilization of Egypt thrived for several millennia. What explains the persistence of irrigated farming in Egypt in the face of its demise in Mesopotamia? The answer lies in the different soil and water regimes of the two lands. Neither clogging by silt nor poisoning by salt was as severe along the Nile as in the Tigris-Euphrates plain.

The silt of Egypt is brought by the Blue Nile from the volcanic highlands of Ethiopia, and it is mixed with the organic matter brought by the White Nile from its swampy sources. It was not so excessive as to choke the irrigation canals, yet was fertile enough to add nutrients to the fields and nourish their crops. Whereas in Mesopotamia the inundation usually comes in the spring, and summer evaporation tends to make the soil saline, the Nile rises in the late summer and crests in autumn. So in Egypt the inundation comes at a more favorable time: after the summer heat has killed the weeds and aerated the soil, just in time for the pre-winter planting.

The narrow floodplain of the Nile (except in the Delta) precluded the widespread rise of the water table. Over most of its length, the Nile lies below the level of the adjacent land. When the river crested and inundated the land, the seepage naturally raised the water table. As the river receded and its water level dropped, it pulled the water table down after it. The all-important annual pulsation of the river and the associated fluctuation of the water table under a free-draining floodplain created an automatically repeating self-flushing cycle by which the salts were leached from the irrigated land and carried away by the Nile itself (Hillel, 1994).

Unfortunately, the soil of Egypt—famous for its durability and productivity in ancient times—is now threatened with degradation. The Aswan High Dam (completed in 1970) has blocked the fertile silt that had formerly been delivered by the Nile. The river itself, now running clear of silt, has increased its erosivity and has been scouring its own banks. And along the estuaries of the Delta there is no more deposition, so the coast has been subject to progressive erosion and to intrusion of sea water (a process likely to worsen as global warming causes the sea level to rise). Finally, the artificial maintenance of a nearly constant water level in the river, necessary to allow year-round irrigation and successive cropping, has raised the water table. So Egypt is now subject to the maladies of waterlogging and salination (to which it had for so long seemed immune) and must invest in the installation of extensive groundwater drainage systems to prevent soil degradation.

Modern Irrigation Methods

In recent decades, revolutionary developments have taken place in the science and art of irrigation. A more comprehensive understanding has evolved regarding the soil-crop-water regime as affected by climatic, physiological, and soil factors. These conceptual developments have led to technical innovations in water control that have made possible the maintenance of near-optimal moisture and nutrient conditions throughout the growing season.

Foremost among these innovations are techniques for high-frequency, low-volume, partial-area applications of water and of nutrients directly into the root zone at rates calibrated to satisfy crop needs. Properly applied, new irrigation methods can raise yields while minimizing waste (by runoff, evaporation, and excessive seepage), reducing drainage requirements, and promoting the integration of irrigation with essential concurrent operations such as fertilization and pest control. The use of brackish water has become more feasible, as has the utilization of sandy, stony, and steep lands previously considered unirrigable. Additional potential benefits include increased crop diversification and cropping intensity; i.e., the number of crops that can be grown in succession each year.

The traditional method of irrigation consisted of flooding the land to some depth with a large volume of water so as to saturate the soil completely, then waiting some days or weeks until the moisture stored in the soil was nearly depleted before flooding the land once again. In this low-frequency, high-volume, total-area pattern of irrigation, the typical cycles consist of periods of excessive soil moisture alternating with periods of insufficiency. Optimal conditions occur only briefly in transition from one extreme to the other.

In contrast, the newer irrigation methods are designed to apply a small, measured volume of water at frequent intervals precisely to where the roots are concentrated. The aim is to reduce fluctuations in the moisture content of the root zone by maintaining optimal (moist but unsaturated) conditions continuously, without subjecting the crop either to oxygen stress from excess moisture or water stress from lack of moisture. Moreover, applying the water at spatially discrete locations, or even below the surface, has the effect of keeping much of the surface dry, thus helping not only to reduce evaporation but also to suppress proliferation of weeds.

Since the high-frequency irrigation systems can be adjusted to supply water at very nearly the exact rate required by the crop, the irrigator no longer needs to depend on the soil's ability to store water during long intervals between irrigations. Hence, water storage properties of the soil, once considered essential, are no longer decisive in determining whether a soil is irrigable. New lands, traditionally believed to be unsuited for irrigation, can now be brought into production. Examples are coarse sands and gravels, where moisture storage capacity is very low and where the conveyance and spreading of water by surface flooding would cause too much seepage.

Of particular interest is the method of *drip irrigation* (also called *trickle irrigation*) and its many variants, such as *microsprayer irrigation* or *spitter irrigation*. Collectively called *microirrigation*, these methods have been gaining acceptance in many areas. The idea of applying water slowly, literally drop by drop, at a rate that is continuously absorbed by the soil's root zone, is not an entirely new notion. What has made it practical is the development of low-cost weathering-resistant plastic tubing and variously designed emitter fittings. System assemblies are now available that are capable of maintaining sufficient pressure in thin lateral tubes to ensure uniform discharge throughout the field, as well as ensuring a controlled rate of drip or spray discharge through the narrow orifice emitters, with a minimum of clogging. A variant of the system is the subsurface placement of a perforated or porous tube that can ooze water continuously with practically no loss due to evaporation.

The application systems described have been supplemented by ancillary equipment such as filters, timing or metering valves (enabling the irrigator to predetermine the quantity, duration, and frequency of irrigation water applied), and even equipment to inject fertilizers and pesticides into the water supply. Mechanical clogging by suspended particles in the water supply can be prevented by proper filtration, whereas chemical and biological clogging by precipitating salts and algae can be reduced by slight acidification and algicidal treatment of the water. Numerous trials in varied locations have resulted in increased yields of both orchard and field crops, perennial as well as annual, particularly in adverse conditions of soil, water, and climate. Drip irrigation has also been applied widely for greenhouses and gardens, and lends itself readily to labor-saving automation.

A variant of high-frequency irrigation applicable to large-scale mechanized farms is the so-called *center-pivot irrigation system*. It consists of a central vertical pipe that delivers water to a horizontal rotating boom fitted with sprinklers or drippers. The boom circles continuously, and can thus irrigate a large area (scores of hectares) automatically. The speed of rotation can be adjusted so that every spot of land (and every plant) receives a small increment

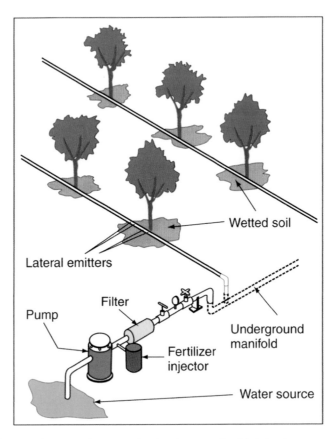

Fig. 13.5. Partial-area wetting around orchard trees under drip irrigation.

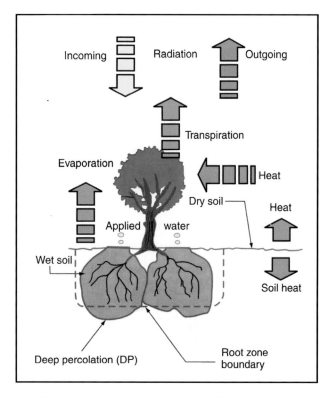

Fig. 13.6. Radiation and water balances on a plant under localized irrigation.

of irrigation at a high frequency; e.g., several times each day throughout the growing season. The application rate can be adjusted according to the variable (weather-determined) evaporative demand and to the stage of crop growth. The circular pattern of irrigation can be seen from the air by airline passengers flying over large sections of the US Great Plains.

WATER-USE EFFICIENCY AND WATER CONSERVATION

Any concept of efficiency is a measure of the output obtainable from a given input. *Irrigation efficiency* or *water-use efficiency* can be defined in different ways, however, depending on the nature of the inputs and outputs considered. For example, one can define as an economic criterion of efficiency the financial return in relation to the money invested in the installation and operation of a water-supply system. The difficulty is that costs and prices fluctuate from year to year and vary widely from place to place, so they are not universally comparable. Another criterion for the relative merit of an irrigation system is an agronomic one—namely, the added yield resulting from irrigation per unit of land area, or per unit amount of water applied.

A widely used expression of efficiency is the *crop water-use efficiency*, which is defined as the amount of vegetative dry matter produced per unit volume of water taken up by the crop from the soil. Because most of the water taken up by plants in the field is transpired (in arid regions, as much as 99%), while generally only a small amount is retained, the plant water-use efficiency is in effect the reciprocal of what has long been known as the *transpiration ratio*, defined as the mass of water transpired per unit mass of dry matter produced by the plants.

What we refer to as *technical efficiency* is what irrigation engineers call *irrigation efficiency*. It is generally defined as the net amount of water added to the root zone divided by the amount of water taken from some source. As such, this criterion of efficiency can be applied to large regional projects, to individual farms, or to specific fields. In each case, the difference between the amount withdrawn from the source and the net amount of water added to the root zone represents the loss incurred in conveyance and distribution.

In practice, many of the older irrigation projects have long operated in an inherently inefficient way. In many surface-irrigated (or flood-irrigated) schemes, water is delivered to the farm at a fixed interval for a fixed period, and charges are assessed per delivery regardless of the actual amount used. Under these conditions, farmers tend to take as much water as they can while they can. This often results in overirrigation, which not only wastes water but also causes problems connected with the disposal of return flows, waterlogging of soils, leaching of nutrients, and excessive elevation of the water table, requiring expensive drainage works.

Particularly difficult to change are management practices that lead to deliberate waste, not because of lack of knowledge or technical means to avoid waste but simply because it appears more convenient or economical in the short run to waste water rather than to apply practices of water conservation. Such situations typically occur when the price of irrigation water is lower than the cost of labor or the equipment needs to avoid overirrigation. Very often the price of water does not reflect the true cost of supplying it but is kept deliberately low by government subsidy, which can be self-defeating.

The agronomic efficiency of water use (WUE_{ag}) can be defined as follows:

$$WUE_{ag} = P/W$$

where P is crop production, either in terms of total dry matter or of the marketable product, and W is the volume of water applied.

Since only a fraction of the applied water is actually absorbed and utilized by the crop, we need to consider the various components of the denominator W, as follows:

$$W = R + D + E_d + E_s + T_w + T_c$$

Here, R is the volume of water lost by runoff from the field, D is the volume drained out of the root zone by deep percolation, E_d is the volume lost by evaporation during delivery and application of water to the field, E_s is the volume evaporated from the soil, T_w is the volume transpired by weeds, and T_c is the volume transpired by the crop. All these volumes pertain to the same unit area and time period.

Clearly, water-use efficiency can be maximized by increasing crop yields P and by reducing the various components of water use W. Crop yields can be increased by using high-potential varieties that are well adapted to local soil and climate, and by optimizing agronomic practices. The latter include proper timing and performance of planting and harvesting, tillage, fertilization, and pest control. Of the items included in the term W, the one that generally cannot, and should not, be reduced is the transpiration by the crop. In the open field, little can be done to limit transpiration without curtailing growth and yield. That is because the unrestricted absorption of carbon dioxide, required in photosynthesis, depends on the stomates remaining open, a condition that allows unrestricted transpiration as well. So, in effect, photosynthesis and transpiration are coupled processes.

The greatest promise for increasing water-use efficiency appears to be in allowing the crop to transpire freely at the maximum climatic limit by preventing any possible water shortages, while avoiding waste of water and obviating all other environmental constraints, resulting in attainment of the fullest possible productive potential of the crop. This is particularly important in the case of superior varieties that can achieve their potential yields only if water stress is prevented and such other factors as soil fertility, temperature, aeration, and soil structure are optimized as well.

Plant diseases and pests, as well as competing weeds and deficient fertility of the soil, may depress yields without a proportionate decrease in water use. All management practices can thus influence water-use efficiency, and none can be considered in isolation from the others.

Given the expected increase in the world's population and the necessity to alleviate human deprivation while maintaining biodiversity and environmental quality, and given the limited availability of appropriate land and water resources, there is an imperative need to improve the efficiency and ensure the sustainability of soil management.

ENGINEERING APPLICATIONS OF SOILS

In addition to its role in agriculture, the soil is used for a variety of purposes in engineering practice. The soil serves as a foundation for many structures, including roads, dams, and houses. The soil also serves as a building material—for example, in the manufacture of bricks and the construction of earthen dykes and adobe houses. Dry soil is a good thermal insulator. Ceramic materials are made of fired clay, in the making of which the particles are fused under heat, and the mass hardens and becomes relatively impermeable.

The tendency of some clay types, especially smectite, to expand when it imbibes water and to contract when it dries can cause buildings, roadways, and pipelines to subside and even to shatter. To avoid failure, structures established over expansive clay must be based on especially strong foundations resting on deep piles or underlying solid bedrock. In some cases, entire cities that are built over saturated clay (Bangkok, Houston, and Mexico City are prime examples) may subside gradually due to the consolidation of the clay underneath.

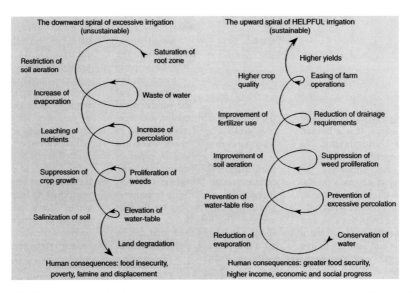

Fig. 13.7.   Inefficient irrigation, downward spriral (a) vs. efficient irrigation, upward spiral (b).

Engineers who deal with the soil as a building material or as a foundation for buildings must consider the mechanical properties of the soil in various states of wetness and under various stresses. The concept of *soil strength* expresses a soil body's ability to bear loads and to withstand compressive or shearing stresses without failing; i.e., without shattering or collapsing. Soil strength is generally a function of the soil's degree of compaction, or its bulk density; i.e., the mass of dry soil per unit bulk volume. To make the soil stronger, therefore, engineers usually try to compact it by means of machines that can knead and compress the soil to a high bulk density. An important consideration is the moisture content of the soil at the time of compaction. Each soil has an optimal moisture content at which a given compactive effort can produce the maximal density and strength. A soil that is too dry will pulverize under stress, and a soil that is too wet will deform when kneaded (as does dough), but in neither state will the soil compress and become stronger. Hence, an intermediate condition is desired at which the soil is moist but not saturated and therefore most amenable to compression.

An important use of soil is in the disposal of wastes. Domestic sewage (including garbage and human wastes), industrial wastes (which may be toxic), surplus building materials, and even spent medical supplies are commonly applied to so-called "landfills". In many cases, the soil's capacity to decompose or immobilize such materials may indeed render them harmless. In too many cases, however, the soil's limited capacities are exceeded, so that harmful agents may leak into the larger environment. The best facilities are designed and controlled landfills that are isolated from the environment by means of compacted clay layers, durable plastic linings, or concrete structures designed to minimize and control the venting of gases to the atmosphere and the leaching of liquids to groundwater and surface streams.

BOX 13.5   Optimizing Rainwater Utilization in Dryland Farming

Improving water-use efficiency in dryland farming requires measures to increase infiltration (avoiding runoff losses) and to prevent water losses. Such measures include the following:

1. Maintaining a well-structured, aggregated, and porous topsoil, so as to prevent surface crusting and runoff;
2. Keeping a mulch cover (consisting of plant residues) on the soil surface to shield the soil surface against the aggregate-slaking impact of striking raindrops;
3. Terracing and contouring cultivation to facilitate absorption of rainfall and prevention of runoff;
4. Avoiding mechanical compaction so as to enhance infiltration and prevent runoff losses;
5. Fallowing the land periodically to collect rainwater and store it in the soil for the subsequent use of a crop;
6. Minimizing surface evaporation of soil moisture by judicious tillage and especially by means of maintaining a diffusion barrier over the surface; e.g., a straw mulch;
7. Eradicating transpiring weeds to prevent losses of moisture from deeper layers of the soil;
8. Enhancing rainwater supply by means of water-harvesting; i.e., inducing and collecting runoff from adjacent slopes and directing it to planted plots;
9. Planting and fertilizing suitable (drought resistant, high yield potential) crops at optimal timing to ensure germination and establishment and to utilize seasonal rains; and
10. Establishing vegetated shelter belts or mechanical barriers (perpendicular to prevailing wind direction) to reduce wind speed and thereby lower potential evaporation.

BOX 13.6   Management of Urban Soils

Especially sensitive are the artificial environments within cities, subject as they are to the so-called "urban heat island" effect, as well as to the urban air's gaseous pollutants and particulates (including dust, soot, and smoke). Urban soils may be artificial media, often mixed with debris of building materials and other wastes (which may be acidic, alkaline, or otherwise toxic). Depending on how urban soils are treated, they may contain increased or decreased amounts of organic matter. In the United States, conversion of agricultural or naturally vegetated land to urban use is proceeding at an accelerating pace. Between 1980 and 2000, land devoted to urban uses grew by more than 34%, a rate 40% faster than the rate of population growth.

Soil organic carbon densities vary widely among different types of urban soils, land uses, and land covers. Soils of residential lawns have the highest content of carbon, whereas soils in parks and recreation grounds contain less of it. Some soils, particularly waste-disposal sites, may contain residues that inhibit plant growth and may, when trampled, contribute to the dust content of the urban air. Properly vegetated and treated, urban soils in parks, streets, residential yards, and even rooftops, can enhance the aesthetic appearance of neighbourhoods, improve the quality of urban air, moderate climatic extremes, help in the control and disposal of storm-water drainage, and even play a positive role in the absorption of carbon dioxide.

14. SOIL EROSION AND CONSERVATION

Thou shalt inherit the Holy Earth
as a faithful servant,
safeguard thy fields from erosion.
Walter Clay Lowdermilk
"An Eleventh Commandment"

GEOLOGIC AND ACCELERATED EROSION

A distinction should be made at the outset between geologic erosion and accelerated (anthropogenic) erosion. The former is a natural and inevitable process that takes place continually and universally. This slow and inexorable process is, however, greatly accelerated by human activity. Denudation of land by deforestation, followed by intensive grazing as well as by cultivations that repeatedly pulverize the soil, induce several modes of soil degradation. Among these are decomposition without replenishment of organic matter, depletion of nutrients, breakdown of aggregates and of structural stability, compaction and crust formation, enhanced runoff, and topsoil erosion by water and wind.

Of these interrelated processes, the most ultimately destructive is topsoil erosion. It results not only in loss of soil productivity on sites where it occurs but also in the off-site deposition of sediments that may pollute surface and underground water resources as well as clog streams, reservoirs, and estuaries. Soil erosion also removes agricultural chemicals (fertilizers and pesticides) from the fields where they had been applied and spreads them in the open environment, where they may disrupt biotic communities and do much harm to non-target organisms. The loss of topsoil to erosion requires farmers to use greater amounts of chemical fertilizers to compensate for the reduced natural

197

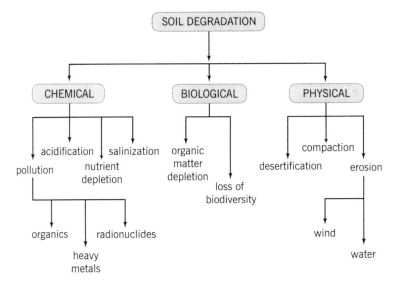

Fig. 14.1.   The various forms of soil degradation. (Dubbin, 2001).

fertility. The increased application of such chemicals itself exacerbates the hazards of groundwater and surface-water pollution.

Erosion by water is the main hazard in relatively humid areas, whereas erosion by wind is a serious hazard in arid conditions. Both processes typically occur in three stages: (1) detachment of particles from the soil body that is their natural domain; (2) transport of the detached soil particles as they are suspended in flowing water or blowing wind; and (3) deposition of the transported particles—a stage called *sedimentation*.

Erosion by water or wind is only one form of soil degradation. Other forms include compaction, crusting, loss of organic matter and of biodiversity, leaching of nutrients, acidification, alkalization, salination, and accumulation of toxic elements or compounds. Of these, however, erosion is generally the most severe, simply because it is irreversible.

WATER EROSION

In humid regions, water rather than wind is the most likely agent of soil erosion. Raindrops seem innocuous when they fall on our heads, but when they strike a bare soil surface, they act like myriads of miniature bombs, each detaching soil particles and gnawing at the soil until they create a layer of dispersed mud over the surface that clogs the pores there with sediment, reducing infiltration and initiating runoff. After the rainstorm ends, that thin layer of mud dries to form a surface crust that inhibits soil aeration and blocks the emergence of germinating seedlings.

To the erosive power of striking raindrops is added the erosive power of running water once surface runoff begins. That happens when the infiltration rate into the soil falls below the rate of rainfall. The excess of rainfall over infiltration ponds over the surface if it is horizontal, or runs off the surface if it is sloping. The erosive power of flowing water depends on the speed of flow, which in turn is a function of the steepness and length of the slope. Surface runoff may start as gentle sheet flow, but as it accelerates, its erosivity increases and it soon begins to scour the soil and form rills, which—if the slope is long and steep enough, and if the soil is loose (i.e., erodible) enough—may converge to form deep gullies.

Soils composed of expansive clay tend to form cracks when they dry, and the cracks provide pathways for rapid entry of water in the early stages of infiltration. The concentration of water in such cracks may destabilize sloping ground along banks of streams and along roadbeds and may cause large blocks of soil to slump or collapse. On flat ground, however, continuing infiltration causes the clay to swell and the cracks to seal, only to re-open during the next period of drying. The repeated heaving and churning of the soil mass can disrupt earthen structures such as dams and roadbeds and undermine buildings.

In water erosion, detachment of soil particles generally occurs under the impact of striking raindrops or the scouring action of water flowing over the soil surface. As it runs downslope, that overland flow, also called *runoff*, carries the detached particles in suspension. When the running water finally comes to rest in a low-lying area, it deposits its suspended load, known as *fluvial sediment*.

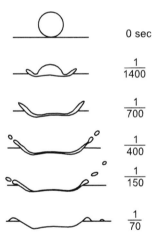

Fig. 14.2.   Impact of raindrops on an erodible soil surface.

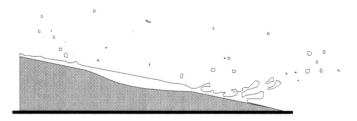

Fig. 14.3.   Initiation of downslope erosion under raindrop impact.

Initial particle detachment (often called "splash") depends on rainfall intensity (the amount of water falling on a unit area in a unit of time, generally expressed as millimeters per hour), rainfall duration, the size distribution of raindrops and their terminal velocities, the direction and steepness of slope, wind, soil conditions (texture, looseness, size and stability of aggregates, and roughness of the surface), and the possible presence of impediments to splash, such as vegetation, litter, and gravel.

A typical raindrop is subject to gravitational acceleration as it falls. As it gains speed, however, the resistance of the air also increases, until it becomes equal to the accelerating force. Thereafter, the drop ceases to accelerate but continues to fall at a constant speed, called the *terminal velocity*. That velocity depends on each drop's mass. The erosive power of rainfall is related to the cumulative kinetic energy of the raindrops, which, for each drop, is proportional to the mass and to the velocity squared, $E_k = mv^2$. Raindrops with a diameter as large as 5 mm may strike the soil at a speed of nearly 10 m/sec. The abrupt collision of such drops with the soil is especially erosive. Different regions are characterized by different rainfall patterns. Rainstorms in some tropical regions tend to be more intense than in temperate regions.

The rates of detachment and transport depend on several interacting variables that are difficult to quantify. In the first place, particle detachment is initiated by striking raindrops, the kinetic energy of which depends on their sizes. That energy, however, is partly absorbed and moderated whenever the soil surface becomes ponded-over by a layer of water resulting from the excess of rainfall rate over the infiltration rate. Detachment is also due to the shearing and lifting forces generated by water flowing over the soil surface, whether that surface is relatively smooth or roughened by protruding particles. Secondarily, the rate of detachment is influenced by the sizes of the grains and the strength of their attachment in aggregates. Generally, the smallest and lightest of the particles (e.g., clay) move first, provided they are not physically or physico-chemically bonded (or "cemented") together in stable flocs and aggregates.

The preceding description is a simplified depiction of the processes of particle detachment and transport, which are in reality very complex. Among the factors that complicate those processes are the variable sizes, shapes, spatial distributions, and bondings of the soil particles, as well as the time-variable and space-variable permeability, roughness, slope angle, and slope length of the soil surface. The physical condition of the soil itself changes during the course of a rainstorm and during a succession of rainstorms in any given season.

Early investigators of water erosion made distinctions among various types or stages of erosion, such as *sheet erosion, rill erosion, gully erosion*, and *stream-channel erosion*. *Sheet erosion* was defined as the uniform removal of thin layers of soil from a relatively smooth slope, carried out by the areally distributed, rather than concentrated, runoff over the entire soil surface. This is an idealized concept that rarely occurs as such. In actuality, the erosion process is hardly ever uniform, and very soon after it begins, the sediment-carrying runoff tends to concentrate in small rills that wend their way downslope. Being small and shallow, these rills are hardly noticeable at first and are easily obliterated by subsequent cultivation. However, as the process continues and is repeated over the course of successive rainstorms, the rills gradually become deeper and wider. They scour and cut into the surface, forming distinct gullies that can no longer be obscured by conventional tillage and that converge downslope to form rivulets. As gullies cut further into the soil, they may reach a tight and practically impervious subsoil that resists further downward scouring. Thereafter, gully erosion continues mainly along the heads and the slumping sides of the gullies.

The amount of erosion depends as much on the erodibility of the soil as it does on the erosivity of the rain. *Erodibility* refers to the vulnerability of the soil to the erosive action of rain and of running water. It depends on the soil's texture and structure. Thus, a dispersed clay is more erodible than a flocculated clay. Similarly, a soil bared and pulverized by excessive tillage or trampled by animals and vehicles is more erodible than an undisturbed and mulch-protected soil. Other relevant factors are the slope and the type and density of vegetative cover.

The following equation is known as the *universal soil loss equation*:

$$A = R\,K\,LS\,C\,P$$

In this empirical expression, A is the average annual soil loss in metric tons per hectare, R the rainfall erosivity factor (depending on its intensity, quantity, and duration), K is the soil erodibility factor (a measure of the soil's susceptibility to erosion), LS is the topographic factor (combining the effect of slope length and steepness), C is the surface cover factor (depending on whether the soil is vegetated, mulched, or bare), and P is the management factor (depending on how the soil is cultivated). The advantage of the universal soil loss equation is

BOX 14.1 **Plato on Soil Degradation**

The Greek philosopher Plato (427–347 B.C.E.), who founded the first so-called Academy in Athens, in one of his Dialogues had Critias bewail the degradation of soil that had already taken place in Greece even two and a half millennia ago.

"What now remains of the formerly rich land is like the skeleton of a sick man, with all the fat and soft earth having wasted away and only the bare framework remaining. Formerly, many of the mountains were arable. The valleys that were full of rich soil are now marshes. Hills that were once covered with forests and produced abundant pasture now produce only food for bees. Once the land was enriched by early rains, for they were not lost as they are now by flowing from the bare land into the sea. The soil was deep, it absorbed and kept the water in the loamy soil, and the water that soaked into the hills fed springs and running streams everywhere. Now the abandoned shrines at spots where formerly were springs attest that our depiction of the land is true."

Has the picture changed? Yes, it has gotten worse. What took place in specific areas is now a global occurrence. The early astronauts described what they saw as they orbited the earth. They noted widespread forest fires in tropical areas and dust clouds in over-grazed semiarid areas. The island of Madagascar, for example, appeared like a wounded giant bleeding into the sea, its red soils raked off the denuded slopes by monsoonal rains and carried away by streams into the ocean.

its simplicity. Its disadvantage is that it may be too simplistic. To be realistic, it must be calibrated in local conditions and based on actual measurements of the variables involved. In reality, soil erosion is not a steady, easily predictable process. Much of it occurs episodically, in rare but violent events. A single torrential rainstorm striking the soil just when its surface is bare and pulverized can cause more soil loss in a few hours than can a season's "normal" rainfall in a vegetated field.

Effective control of erosion by water consists of minimizing the impact of raindrops and the velocity of running water on the soil surface. This includes enhancing infiltrability and surface storage, improving soil structure, protecting the surface by a cover crop or a mulch to prevent raindrops from striking the exposed soil, minimizing cultivation and performing it on the contour rather than up and down the slope, and avoiding compaction and excessive pulverization. An ancient and still common practice of soil conservation is the shaping of sloping land by means of terraces or contour strips to reduce the inclination and the length of the slope segment, thereby checking the velocity of running water.

BOX 14.2   Demonstrating the Significance of Soil Erosion

Two early leaders of the US Soil Conservation Service, Hugh H. Bennett and Walter Clay Lowdermilk, who surveyed the global dimensions of the soil erosion problem, wrote in the 1938 *Yearbook of Agriculture*: "Soil erosion is as old as farming. It began when the first heavy rain struck the first furrow turned by a crude implement of tillage in the hands of prehistoric man. It has been going on ever since, wherever man's culture of the earth has bared the soil to rain and wind."

When asked to testify on soil erosion before a committee of Congress, the two of them, without a word, placed a thick towel on the committee's polished table and poured a large cup of water onto it. The towel soaked up the water. Next they removed the wet towel and, still saying nothing to the puzzled members of the committee, poured a second cupful on the bare table. The water splashed and trickled off the table and onto the laps of the distinguished committee members. Every one of them then understood the dire consequences of removing the soil cover from the land.

BOX 14.3   Mulching

The term *mulching* refers to the practice of leaving crop residues on the surface rather than burning them, mixing them in the soil mechanically, or plowing them under. The presence of a mulch anchored in the topsoil and spread over its surface protects the soil against desiccation and deflation by wind, as well as against the beating action of raindrops and their effect of slaking and crusting the surface. The surface mulch reduces runoff and erosion, as well as evaporation.

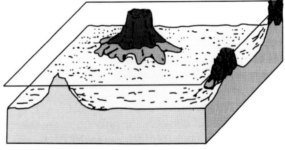

A. Sheet erosion

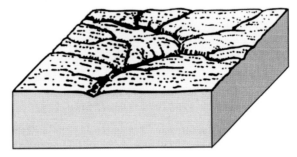

B. Rill erosion

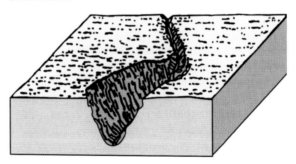

C. Gully erosion

Fig. 14.4. Sheet, rill, and gully erosion (FAO, 1987).

WIND EROSION

Bare soil surfaces, especially when dry and loose, are vulnerable to erosion by wind. The process of wind erosion, like that of water erosion, consists of three more-or-less distinct phases: detachment and lifting of particles; transportation; and deposition. As in the case of water erosion, the entire process can be regarded as an interaction of the *erosivity* of the fluid medium (in this case, the flowing air) and the *erodibility* of the soil surface.

The detachment and uplifting of particles result from the effects of wind gusts and air turbulence in the lower layer of the atmosphere on the soil surface. In general, the minimal wind speed, measured at a height of 0.3 m above the ground surface, needed to initiate wind erosion has been found to be about

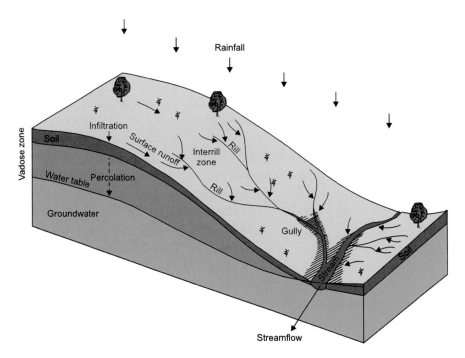

Fig. 14.5. A denuded landscape subject to rill and gully erosion.

5 m/sec. The initiation of wind erosion is strongly affected by the microtopography of the soil surface, especially by the presence of protruding hillocks or hummocks. It is also enhanced by the midday heating of the dry soil surface, which creates conditions of aerodynamic instability as the air that is warmed by the soil becomes buoyant and tends to rise. In desert areas, the combination of this buoyancy and the turbulent wind gusts striking the ground promotes the uplift of soil particles and the appearance of spiral columns known as *dust devils*. Such "mini-tornados" may rise to heights of scores of meters and seem to flutter over the desert floor like whimsical dancers.

In wind erosion, air gusts pick up masses of particles from the surface, an action called *deflation*, and then convey this material, called *aeolian dust*, over some distance before depositing it. That distance can be great indeed. (The circulation of some wind-blown dust has been shown to be global: Dust particles originating in the Sahara have even been identified in Hawaii!) The proneness of the soil to wind erosion depends on surface dryness, roughness, soil texture, and whether the particles are loose or bound in aggregates.

Three types of soil movement by wind can be recognized: *surface creep; saltation*; and *suspension. Surface creep* is the rolling or sweeping motion of relatively large particles (having a diameter of 0.5 mm or more) that occurs within a few centimeters over the ground surface. *Saltation* is the jerky, kangaroo-like motion of finer sand particles (0.1–0.5 mm) that follow distinct trajectories. Particles are lifted and transported over a distance of some meters and then—as they strike the ground obliquely—they bounce away another particle, in a repeated sequence of hop-skip-bounce events. Most saltation occurs within a half meter of the soil surface. Still finer particles (0.02–0.1 mm), once lifted off the surface,

Fig. 14.6. Effect of a mulch on protecting the soil surface against rainsplash. Top: surface covered by plant residues. Bottom: bare surface.

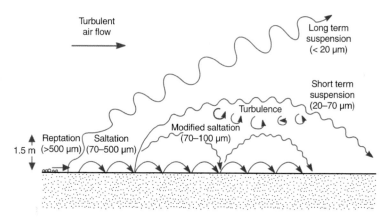

Fig. 14.7. Wind sweeping and bouncing particles over a dry surface. (Schultz, 2005).

remain suspended in the air for extended periods. They are often carried aloft and may be transported by air currents to distances of hundreds of kilometers.

The quantity and distance of windborne soil movement depends on the array of particle sizes, the wind speed, and the topography of the land. The capacity of the air to transport particles generally varies as the cube of the wind speed above a threshold value, and inversely as the square root of the particle diameter. Deposition of the sediment takes place where the wind abates so that the gravitational force exceeds the aerodynamic lift. Wind speed diminishes especially on the lee side of clumps of vegetation or other physical barriers; e.g., fences of shelterbelts. Raindrops also bring down dust from the air.

Various measures have been devised to control wind erosion. Among these are the maintenance of a protective sod or mulch over the soil surface, the increase of surface cloddiness and roughness by means of appropriate tillage and shaping of the land, the stabilization of surface-zone soil aggregates, and the use of mechanical or vegetative windbreaks.

BOX 14.4 Soil Erosion in the USA

In the early history of the United States, land seemed plentiful. If an area in the eastern states lost its productivity, farmers could move westward and claim virgin land to cultivate (and, in the process, displace the Native American tribes). As the new settlers moved into the Midwestern Prairies, it seemed that they came upon ever more fertile soils. Eventually, however, as the continent was filled and most of the arable land appropriated, the realization set in that better methods of land management, based on soil conservation and improvement, were needed. That realization became imperative in the 1930s, when the devastating effect of mismanagement became inescapably evident: 24% of the farmland had been severely eroded (having lost half or more of its topsoil) and another 24% was significantly eroded. Only some 28% of the farmland remained apparently unaffected by erosion. The most dramatic manifestation of soil erosion occurred in the spring and summer of 1934, when a series of dust storms deflated the plowed lands of the southern Great Plains and carried great clouds of billowing dust for thousands of kilometers across the continent to the East Coast. (See next box.) That event, known as the great American Dust Bowl, is far from unique. Similar instances of widespread wind erosion have recurred in Central Asia, in the Middle East, and, on a vast scale, in the Sahel region of sub-Saharan Africa. Wind erosion is most likely to occur in semiarid areas, when the soil surface, bared of protective vegetation and pulverized by tillage or overgrazing, dries out during a drought and is deflated by strong winds.

BOX 14.5 The "Dust Bowl" of the 1930s in the US

A spectacular example of wind erosion occurred in the 1930s in America's Southern Great Plains. Known as the Dust Bowl, it resulted from the introduction of the plow into the vast, semiarid, wind-swept grasslands of that region. During the first three decades of the twentieth century, the region enjoyed a wet period. The prairie was green, and when plowed and planted, it yielded bountiful crops of wheat. But rainfall there is unreliable. In Oklahoma's Cimarron County, for example, the annual rainfall averaged about 500 mm between 1900 and 1930, and as much as 700 mm between 1914 and 1923. Then in 1934 and the following few years, it fell below 350 mm.

During the boom decades, thousands of farmers settled in the region. They broke the sod, planted their seeds in the soft soil, watched the rain-satiated wheat come up, and then harvested its bumper crop. In the Texas Panhandle alone, the area under wheat had expanded from 35,000 to 800,000 hectares from 1909 to 1929. This enormous change in the land's surface was facilitated by the mass production of tractors, which were then leading Dust Bowl villains. They were "snub-nosed monsters," wrote John Steinbeck in *The Grapes of Wrath*, "raising dust and sticking their snouts into it."

Eventually, the copious rains ceased and the drought began. The year 1934 was particularly hot and dry, and the land lay bare and parched under the merciless sun, its pulverized soil prey to the whipping winds. The winds swept up the loose soil as if it were talcum powder and lifted it in billowing red-brown clouds that eclipsed the sun and covered fences, clogged houses, and choked animals and people. So a great exodus began—40% of the people of Cimarron County, for example, abandoned their homes and moved out of the region. The effects of the Dust Bowl were not confined to that region. The dust wafted aloft and drifted across the entire continent. On May 12, 1934, *The New York Times* reported that the city was obscured in a half-light "... and much of the dust lodged itself in the eyes and throats of weeping and coughing New Yorkers." The Eastern Seaboard of the US was blanketed in a heavy fog composed of millions of tons of the rich topsoil swept up into the continental jet stream. Even ships far out in the Atlantic were showered with Great Plains dust.

The Dust Bowl of the 1930s is not just a thing of the past. It is repeated from time to time on a vast scale in such regions as the Sahel in Africa and in the Aral Sea region of central Asia. The process by which a semiarid region is denuded of vegetation and its soil destabilized and rendered vulnerable to erosion is now called *desertification*.

DESERTIFICATION

Terrestrial ecosystems in semiarid and arid regions around the world appear to be undergoing processes of degradation affecting biotic communities and the soils upon which they depend. Especially vulnerable are regions in which the ratio of total annual precipitation to potential evapotranspiration averages below 0.6. Such regions constitute some 40% of the global terrestrial area. They include northern, southwestern, and parts of eastern Africa; southwestern and central Asia; northwestern India and Pakistan; southwestern USA and Mexico; western South America; and much of Australia. These regions are home to an estimated sixth of the world's population.

Desertification is a term used to cover a wide variety of interactive phenomena—both natural and anthropogenic—affecting the actual and potential biological and agricultural productivity of ecosystems in semiarid and arid regions. While there are places where the edge of the desert can be seen encroaching on fertile land, the more pressing problem is the deterioration of land due to human abuse in regions well outside the natural domain of the desert. The latter problem emanates not from the expansion of the desert *per se* but from the centers of population outside the desert, owing to human mismanagement of the land. A vicious cycle has begun in many areas: as the land degrades through misuse, it is worked or grazed ever more intensively, so its degradation is exacerbated; and as returns from "old" land diminish, "new" land is brought under cultivation or under grazing in marginal or even submarginal areas.

As commonly defined, desertification is the process by which an area becomes (or is made to become) desert-like. The word *desert* itself is derived from the Latin *deserere*, meaning "to desert" (to abandon). The clear implication is that a desert is an area too barren and desolate to support human life. An area that was not originally desert may come to resemble a desert if it loses so much of its usable resources that it can no longer provide adequate subsistence to a given number of humans. This is a very qualitative definition, since not all deserts are the same. An area's resemblance to a desert does not make

it a permanent desert if it can recover from its damaged state, either spontaneously (if left alone) or via human-induced remediation.

A typical feature of arid regions is that the mode (the most probable) amount
of annual rainfall is generally less than the mean; i.e., there tend to be more
years with below-average rainfall than years in which the rainfall is above
average, simply because a few unusually rainy years can skew the statistical
mean well above realistic expectations for rainfall in most years. The variability of biologically effective rainfall is yet more pronounced, as years with less
rain are usually characterized by greater evaporative demand, so the moisture
deficit is greater than that indicated by the reduction of rainfall alone. Timing
and distribution of rainfall also play a role. Below-average rainfall, if well distributed, may produce adequate crop yields, whereas average or even above-
average rainfall may fail to produce adequate yields if the rain occurs in just a
few ill-timed storms with long dry periods between them.

In arid and semiarid regions, drought is a constant menace. Its occurrence
is a certainty, sooner or later; only its timing, duration, and severity are ever
in doubt. And it is during a drought that ecosystem degradation in the form
of denudation and soil erosion occur at an accelerated pace, as people try to
survive in a parched habitat by cutting the trees for fuel and browse, and by
animals overgrazing the wilted grass. The topsoil, laid bare and pulverized by
tillage or the trampling of livestock, is then exposed to a greatly increased risk
of wind erosion. When the coveted rains recur, they tend to scour the erodible
soil. Whenever human pressure on the land ceases or is diminished, even a
severely eroded soil may recover gradually. However, on the time-scale of years
to a few decades, especially if overgrazing and overcultivation continue, soil
erosion may become, in effect, irreversible.

CONTROL OF SOIL EROSION

Effective control of soil erosion by water consists of minimizing the impact
of raindrops and the velocity of running water on the soil surface. This task
includes enhancing infiltrability and surface storage, improving soil structure,
protecting the topsoil by a cover crop or a mulch of organic residues (e.g.,
straw) to prevent raindrops from striking the bare surface, minimizing cultivation and performing it on the contour rather than up and down the slope, and
avoiding both compaction and excessive soil pulverization. An ancient and still
common practice of soil conservation is the shaping of sloping land by means
of terraces or contour strips to reduce the inclination of the surface and the
length of slope segment, thereby checking the downhill acceleration of running
water.

Control of wind erosion can be achieved by means of shelter belts, which
are parallel rows of trees or shrubs planted in a direction perpendicular to the
prevailing direction of the wind. Additional measures are ensuring the presence
of a protective vegetative cover or a mulch on the soil surface to keep it from
the direct action of wind; keeping the topsoil in a cloddy rather than dusty
or excessively pulverized state; enhancing soil aggregation by organic-matter
enrichment; and maintaining the topsoil in a moist condition by evaporation
control and, where possible, by light irrigation.

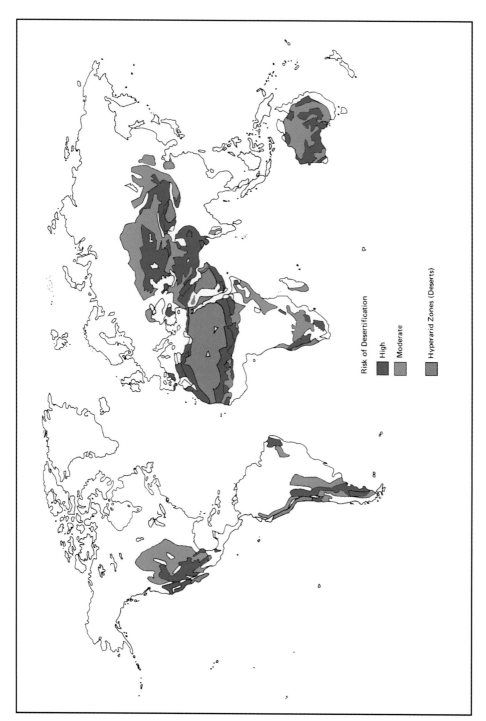

Fig. 14.8. Deserts and areas subject to desertification. (Nebel, 1987).

Risk of Desertification

High

Moderate

Hyperarid Zones (Deserts)

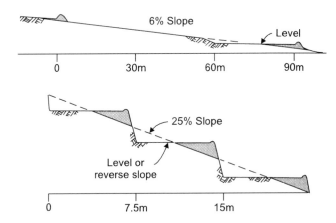

Fig. 14.9.   Terrace shapes on slopes of different steepness.

 BOX 14.6   Soil Erosion and Conservation

Soil erosion is the removal of soil material, generally from the surface, by water (striking raindrops and downslope runoff) or by wind (deflation of loose particles and their transport as dust). Erosion is a natural process that occurs inevitably and continuously. However, its rate can be greatly accelerated by removal of protective vegetation and its residues and by mechanical cultivation or trampling of the soil.

Accelerated erosion may cause great damages to soils and to ecosystems:

1. The stripping away of topsoil results in loss of fertility.
2. Transported water-borne sediment clogs streams, lakes, estuaries, and dams.
3. Sediment may cause rivers to overflow, change course, and inundate adjacent lands.
4. Sediment increases the turbidity of lakes and causes their eutrophication, to the detriment of native aquatic species of fish, etc.
5. The transport of wind-blown sediment may damage crops by abrasion and blockage, cause atmospheric turbidity, partially block sunlight, and affect the respiration of animals, including humans.

 BOX 14.7   Tropical Soils: Challenges and Imperatives

Of all the world's variegated soils, those of the tropics and subtropics are particularly vulnerable to exploitation. Many of them are highly vulnerable to erosion as well as to depletion of organic matter, owing to its rapid rate of decomposition in a warm environment and its unstable storage in light-textured, highly leached soil profiles. Farmers in those regions are typically unable to afford the inputs necessary to sustain the fertility of their soils and to enrich it with organic residues. What residues are available to them are often used as fuel or livestock feed. For these reasons, the fragile soils of the tropics (both those of the humid tropics and those of the semiarid or arid tropics) require special techniques and care in their management. All these facts make it all the more imperative to devote special attention and effort to the amelioration of tropical soils.

15. SOIL POLLUTION AND REMEDIATION

For poisoned air and tortured soil ...
every secret woe the shuddering water saw,
willed and fulfilled by high and low
—let them relearn the Law.
Rudyard Kipling, 1865–1936

LAND DISPOSAL OF WASTES

Modern civilization's profligate use of varied chemicals in industrial, commercial, and domestic activities has led, often inadvertently, to serious contamination of air, water, and soil. Among the chemicals in common use are many that spur food production, alleviate disease, ease and prolong human life, and serve as catalysts for industrial processes. But once their residues leak or are discarded into the environment, some of the chemicals or their derivatives may disrupt entire ecosystems. The wastes produced by our industries include substances that are toxic to many species and hazardous to humans. In addition, household and municipal wastes, including sewage sludge, food-processing remains, domestic garbage, building materials, petroleum products, and other disposable materials too numerous to list, tend to clutter modern life. Their safe disposal has become one of the most pressing environmental issues of our time.

The traditional and time-honored practice of waste disposal is to apply it to the land. The premise underlying this practice is that the soil's evident capacities to retain, filter, decompose, recycle, or immobilize the waste products of the natural biome can be utilized to rid us of our anthropogenic wastes. In many cases, controlled amounts and qualities of wastes can be applied to the land safely in appropriate, protected locations (generally called "landfills"). However, in too many other cases, the amounts and kinds of waste materials currently applied to the land, including highly toxic and persistent chemicals, overwhelm the soil's limited capacities. Instead of serving as the ultimate depository for the wastes, the overloaded soil has become a mere way station, detaining rather than retaining or neutralizing wastes. Worse yet, in many places the soil has been so polluted as to have itself become a source of toxicity; when eventually it releases the products added to it, the soil transmits the pollutants to surface waters and groundwaters, or, via the organisms residing in the soil, to the food chain at large.

In contrast with wars and natural disasters—earthquakes, volcanic eruptions, or hurricanes—which strike swiftly and inflict their punishment suddenly and spectacularly, the unnatural disaster caused by soil-borne toxins is gradual, and, for a time, ambiguous. We can surmise their impacts only by the statistics of probability, and only after they have been in effect for too long to prevent them. In recent decades, more and more communities have discovered the time-delayed effects of toxins originating in waste disposal sites that contain industrial pollutants, including petroleum derivatives, synthetic organic compounds (e.g., pesticides), and heavy metals; e.g., lead, arsenic, cadmium, zinc, and mercury.

Much hazardous waste is still improperly discarded, simply by dumping it wherever it is convenient or profitable to do so. For too long, under the protection of property rights, a person or company could discharge or bury wastes on private land with impunity, or pay others to deposit it on their land. Hardly any consideration was given to the rights of future generations, and none at all to other affected fauna and flora. Moreover, insufficient consideration has been given to the tendency of hazardous substances to spread from their sites of deposition to other domains of the environmental continuum.

Organized land disposal of wastes is usually done in facilities that are often misnamed "sanitary landfills". The older ones were simply unlined and scarcely protected pits or valleys that eventually grew into artificial mounds towering high above the surrounding ground. In recent decades, legal regulations have been promulgated requiring landfills to be underlain, and sometimes overlain, with clay liners. These are layers of clay, generally several meters thick, which are compacted for the purpose of containing the hazardous materials to be dumped into the landfill. Artificial liners of plastic are sometimes added. More modern landfills are equipped with drains and collection tanks, to prevent the noxious chemicals from leaking into the open environment. However, no system of containment is totally impervious, and many liners and drains tend to develop cracks or yield to the corrosive action of potent chemicals. Many "leakproof" landfills can thus become hazardous sites in themselves.

A particular problem in many locations is the disposal of domestic liquid wastes, including human wastes. The average person in the United States uses about 200 liters of water per day for cooking, washing, and sanitary disposal. The wastewater is commonly collected in a septic tank and then passed into a sand filter or an open leaching field. The ability of the soil in the leaching field and its biota to purify the wastes dissolved and suspended in the water depends on the type of soil, its permeability, and its depth to the water table or to bedrock.

The terms *contamination* and *pollution* have often been used interchangeably. The difference in nuance, however, is that *contamination* refers to the addition of potentially harmful substances, whereas *pollution* refers to a condition in which contamination has reached a level such that the medium is actually impaired, especially if it has become toxic to beneficial organisms.

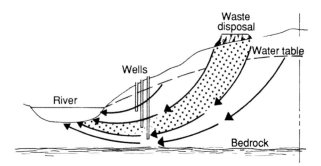

Fig. 15.1. Leaching of contaminants from a waste-disposal site into groundwater.

BOX 15.1 Agriculture as Potential Polluter

Certain agricultural activities may result in environmental pollution. Among such activities are the excessive applications of fertilizers and pesticides. Residues of these activities may persist in soils and migrate to groundwater. Pesticides by their very nature are toxic. They become pollutants when they accumulate and persist in the soil and groundwater, where they can affect non-target organisms, as well as when their residues occur on food. The ideal pesticide is specific to the pest to which it is targeted, and degrades quickly in the environment to a harmless form.

BOX 15.2 Soil as Potential Contaminant

The soil can absorb and degrade many types of potential contaminants, particularly organic wastes, but its capacities are not unlimited. Where its powers of degrading contaminants are exceeded, the soil itself may become a contaminant. When eroded by water and wind, soil material (sediment) and the contaminants it carries can accumulate in streams and ponds, making them turbid and toxic, thus limiting the growth of aquatic plants and fish. Soil material carried in suspension can also clog reservoirs and estuaries.

INORGANIC POLLUTANTS

Among the varied sources of soil contamination are fertilizers, heavy metals (often contained in sewage sludge), solvents, detergents, and accidental spills of various industrial products. Soils can also be contaminated by salts from irrigation water, saline seeps, seawater intrusion along coasts, and deicing treatments of roadways and sidewalks. Such contaminants may accumulate in the soil or leach from it into groundwater, streams, and ponds.

The most hazardous soil pollutants are those discharged from industrial activities such as mining, metal processing, smelting, electroplating, fuel burning, painting, pesticide spraying, fertilizer spreading, sewage sludge application, and numerous other chemical processes. Among the toxic elements discharged by industry are the following: lead, arsenic, cadmium, chromium, mercury, copper, molybdenum, nickel, zinc, and selenium.

Some of the metals mentioned are adsorbed as cations onto negatively charged particles of clay or are associated with organic-matter complexes. Those metals are bonded (and therefore immobilized) more effectively where the soil is well aerated (oxidized) and where it is chemically neutral or slightly alkaline. However, these metals may become quite mobile under acidic and anaerobic (chemically reduced) conditions, especially in sandy soils with limited cation adsorption and low organic matter content.

BOX 15.3   Heavy Metals

Heavy metals are elements with relatively high atomic mass. Some of them (including iron, manganese, copper, nickel, cobalt, zinc, molybdenum, tungsten, and vanadium) function as oxidation-reduction intermediates or as cofactors of enzymes or even as constitutents of vitamins or haemes, so they are needed in small amounts by living species and can be regarded as trace nutrients. Other heave metals (e.g., lead, cadmium, mercury, and silver) have no such functions. In any case, the heavy metals tend to become toxic when present in higher concentrations in soluble forms. Their solubility depends on environmental factors such as pH, redox potential, and the presence of anions and of chelating molecules. Toxic concentrations of these metals do not necessarily result from human-induced pollution, as they may originate from the soil's own parent material. Soils and crops prone to concentrate such toxic factors must be monitored carefully, lest they constitute a hazard to public health.

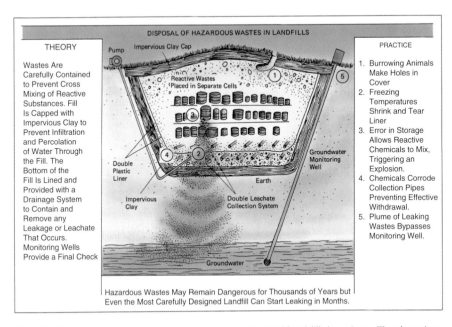

Fig. 15.2.   Disposal of hazardous wastes in a constructed land-fill depository. The depository is lined and capped with impervious clay to prevent infiltration and percolation of water. Controlled drainage is provided to ensure the safe disposal of waste residues. (Nebel, 1987).

Yet another inorganic material that has proven to be highly carcinogenic and environmentally persistent is asbestos. It is a fibrous, flaky material, composed of magnesium silicate, formerly used as an insulator in the construction of many buildings.

In addition to ordinary forms of matter, radioactive elements (e.g., the radionuclides strontium 90, iodine 131, and cesium 137) from military programs and from atomic power plants (e.g., Chernobyl, 1986) have been added to the

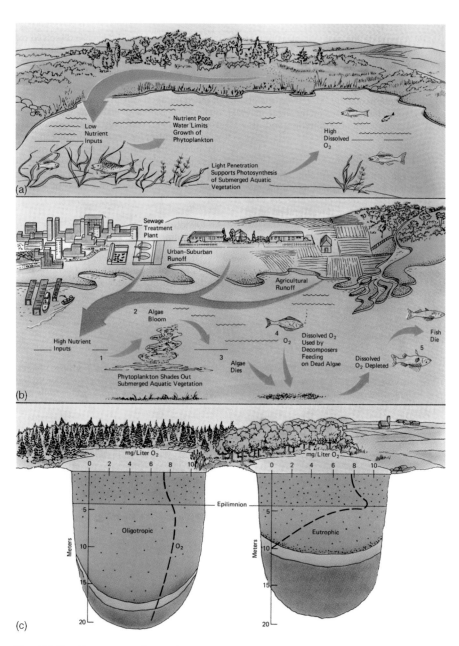

Fig. 15.3. The process of eutrophication (Nebel 1987).

environment here and there, intentionally or otherwise. Depending on the mode of discharge, these (and other) pollutants may merely contaminate the immediate vicinity of industrial complexes, or be carried some distance away by wind or water before coming to rest in the soil. Cesium 137 is adsorbed tightly by clay minerals, which reduce its availability, but strontium 90 can substitute chemically for calcium and thus accumulate in the bones of animals ingesting plants grown in contaminated soil. Many soils naturally contain small amounts of radioactive elements such as uranium, thorium, potassium, radium, and radon. The disposal of radioactive waste from mining activities and the fallout from nuclear tests and power-plant accidents can add greatly to that natural radioactivity.

Some potentially toxic elements (e.g., arsenic, molybdenum, and selenium) may be present in the soil as anions, whose mobility and bioavailability are lower under acidic conditions due to adsorption onto the positive charges that develop in mineral and organic constituents of the soil at low pH values. The metallic element chromium may assume different oxidations states depending on soil aeration. When oxidized, chromate becomes mobile and quite toxic. However, this element is readily immobilized by even small amounts of organic matter.

In some cases, nutrients contained in the soil (either naturally or from fertilizer residues) can become pollutants. Principal sources of phosphorus in the soil are the gradual dissolution of P-containing minerals (e.g., apatite) and the bones of animals. To these sources are added the residues of industrial products (e.g., detergents and certain pesticides). When nutrients, including nitrates and phosphates, drain from the soil and flow into streams and lakes, they may spur the excessive proliferation of aquatic plants and algae. Such growth can deprive the water bodies of sunlight and of oxygen. Indeed, some lakes located in the midst of agricultural districts and urban centers have already, in effect, become lifeless.

Excessive applications of fertilizers and pesticides (including herbicides intended to control weeds, nematocides to suppress crop-damaging nematodes, and insecticides to combat insect pests) may in some instances do more harm than good. Where they leak into groundwater aquifers, remediation may be practically impossible. Several early pesticides (e.g., DDT) were found to produce so much harm to wildlife that they were subsequently banned altogether.

BOX 15.4   Acid Rain

Automobile exhausts and the smokestacks of coal-burning factories spew out smoke that contains sulfur as well as nitrogen oxides. When the released chemicals dissolve in rainwater, they acidify the rain, forming sulfuric and nitric acids. The effects of acid rain can be seen in the corrosion of exposed metals and the weathering of limestone structures, especially of marble statues. Less visible but even more important are the effects on the acidity of the soil and the dissolution and leaching of essential nutrients, including calcium and magnesium. Sulfuric acid can react with some minerals in the soil and in bedrock, especially from marine sedimentary rocks, to release potentially toxic concentrations of metals such as lead, zinc, arsenic, and cadmium. Such metals are commonly present in the sewage released from mining and metal processing industries.

BOX 15.5   Drainage problems in California

Irrigation in the western section of the Central Valley of California is facing a crisis. To avoid salination, the area requires drainage. A mountain range prevents direct conveyance of effluent westward to the Pacific Ocean, so planners intended to direct drainage northward to San Francisco Bay. The first leg of the drainage canal was constructed and terminated at Kesterson Reservoir. This reservoir was to serve as a wildlife habitat and storage facility until the last leg of the drainage canal could be completed. However, the plan to terminate the canal in San Francisco Bay met opposition from Bay-area residents, who feared the bay's contamination by agricultural chemicals.

Just a few years later, biologists monitoring the Kesterson Reservoir noticed that young birds were deformed and died before reaching maturity. Other animals were similarly affected. The culprit was found to be selenium, an element that — ironically — is an essential

nutrient in small amounts, but becomes toxic at higher concentrations It originated in the sedimentary substrata underlying the region, and hence entered the groundwater and the effluent drained from it. The selenium discharged into the open reservoir became progressively concentrated as evaporation took place there. So the region's agricultural planners had to abandon the entire scheme. Now the district is left without an outlet for its drainage, and may be doomed to suffer the fate of other irrigated regions throughout history, unless some other solution is found (and it is likely to be expensive).

Irrigation in California is responsible for the continued existence of another body of water, the Salton Sea, located in the Imperial Valley near the Mexican border. Although repeatedly inundated in past ages, this basin was a dry depression in recent history. In 1904, the Colorado River breached the channel that was built to contain it, and flowed into that depression to form a new fresh-water lake. By 1907, the breach was repaired, and the new lake would normally have evaporated away. Contrary to expectations, the lake has remained, and even grown, since then. It is sustained by the collective agricultural drainage from the Imperial Valley, and sewage effluent from Mexicali. The Imperial Irrigation District is now one of the most intensively irrigated districts in America. The Salton Sea is also used for recreation and is an important wildlife habitat on the Pacific flyway, and it is considered a permanent feature of the valley. However, the salinity of this land-locked lake, lacking any cleansing through-flow (its only outlet being the evaporative sink of the dry desert air) has increased steadily and is now greater than the salinity of ocean water.

BOX 15.6   The shrinking and salination of the Aral Sea

An ecological debacle has occurred in the arid plains of central Asia, a part of the former Soviet Union, as a result of what had earlier seemed to be a shining success of large-scale irrigation development. With its warm, sunny climate, that region was a rich agricultural production center that yielded more than 33 percent of the USSR's fruit, 25 percent of its vegetables, 40 percent of its rice, and as much as 95 percent of its world-leading harvest of cotton.

To provide the water needed for irrigation, engineers diverted the flow of the region's two rivers, the Amu Darya and the Syr Darya, both of which flowed naturally into the Aral Sea. So much of the water was siphoned off that what was the world's fourth largest lake shrank to less than half its original size. The once-thriving fishing industry has been devastated, as fishing villages once located on the shore became stranded 30 to 80 kilometers inland. Moreover, the salinity of the water has risen dramatically.

Behind the receding waterline lie mudflats covered by fluffy salts, which are picked up by the swirling continental winds and float in deadly dust clouds to destroy crops and poison land for hundreds of kilometers around. Compounding the damage of salinity are the residues of agricultural chemicals (fertilizers and pesticides), applied in huge overdoses in an effort to coax the greatest yields from the land in the shortest possible time. The chemicals have seeped into the groundwater and the surface streams, poisoning the only water supply available to the region's population.

Long-proposed schemes to divert the waters of Siberian rivers and to channel them south to the Aral Sea were abandoned after years of controversy over their costs and potential environmental impacts. The more practical approach is to improve the inefficient irrigation system by lining canals, laying pipes in place of open channels, applying water by means of sprinklers or drippers instead of by surface flooding, introducing volume-controlled sluices and valves, and altogether promoting greater efficiency in the distribution and utilization of irrigation water. Equally important is the selection of crops and agronomic methods to conserve water and prevent pollution.

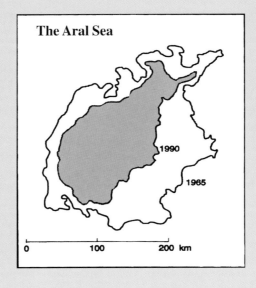

ORGANIC POLLUTANTS

A bewildering array of organic compounds are synthesized by the chemical industry to serve a variety of purposes, including pest control, detergents, solvents, and emulsifiers. Many of these compounds, when introduced into the environment, prove to be quite persistent, and some can be harmful to a variety of organisms.

Pesticides include *insecticides* to eradicate insect pests, *herbicides* to control weed proliferation, and *fungicides* to combat fungal infestations. Among the most recalcitrant pesticides are chlorinated compounds, which can greatly affect non-target organisms. Another class of pesticides, the organophosphates, are highly toxic but degrade more readily. The chemical industry keeps trying to formulate pesticides that are specifically targeted and that degrade rapidly after serving their purpose. Such laudable efforts, however, have met with only partial success and tend on occasion to produce unintended consequences.

Many of the organic compounds applied to the environment end up in the soil, where they are adsorbed by clay minerals or are absorbed by organic matter. These compounds tend to degrade gradually in the soil, chemically or, more specifically, biochemically. Compounds that are highly soluble, however, may be leached from the soil (especially if it is sandy, highly permeable, and with low adsorptive capacity) and may contaminate streams, lakes, or groundwater. Other compounds (e.g., solvents and fuels) may be lost from the soil by volatilization. Residues may remain on or in plants, however, and may well pose a hazard to humans or animals that ingest such plants.

BOX 15.7 Degradable and Undegradable Organic Compounds

Compounds occurring in nature are, as a rule, completely degradable under suitable environmental conditions. However, they must be considered polluting if they occur in high concentrations or if the existing soil conditions (e.g., lack of oxygen) prevents their decomposition.

Artificial compounds (called *xenobiotics*, meaning "foreign to life") may not be degradable under natural conditions. Some of these compounds are largely inert, as are many plastics; while others are toxic. The latter compounds may tend to concentrate progressively along food chains, leading eventually to severe physiological harm even if the concentration in the soil itself is low. An example is the harmful effects of certain PCBs (polychlorobiphenyls) on the eggshells of birds.

Especially insidious are hydrocarbons applied, either purposely or inadvertently, to the soil. Such materials have been leaking into the soil and the underlying groundwater from hundreds of thousands of underground storage tanks over many decades, and their effects are now being noticed in many regions in the US and abroad.

Hydrocarbon fluids (liquids and gases) move through the soil's pores, interact with its aqueous phase and with the surfaces of its solid particles, are directed

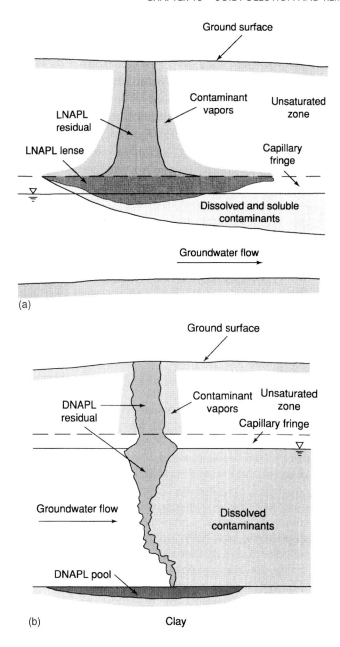

Fig. 15.4. Soil and aquifer pollution by light and dense nonaqueous-phase liquids (LNAPL and DNAPL) (Rubin 2005).

by gravity and the structure of the soil profile, and are acted upon by its biota. The fluids may be miscible or immiscible in water, more or less dense or viscous, and labile or inert. The following is a partial list of processes that may occur as petroleum-derived fluids are introduced (generally inadvertently) to the soil:

1. Volatilization of the lighter constituents, such as low molecular-weight hydrocarbons (e.g., benzene);
2. Runoff over the soil surface, driven by rainstorms, with consequent contamination of surface waters or concentration in surface depressions;
3. Adherence to the soil surface, possibly causing clogging and hydrophobization of the soil;
4. Infiltration into the soil profile;
5. Drawdown and lateral spreading within the soil profile's unsaturated zone, especially over boundaries of less permeable layers;
6. Retention at layer-interfaces within the soil profile;
7. Retention in soil pores and attachment to particles and organic matter;
8. Vapor diffusion within the soil profile;
9. Chromatographic separation of components within the profile, resulting in selective migration of lighter and less viscous (hence more mobile) components;
10. Partial dissolution of soluble or emulsifiable components within the water phase of the soil;
11. Degradation of the invading compounds by either or both chemical and biological processes;
12. Leaching from the soil, driven by or dissolved in the water phase toward the water table;
13. Mounding over the water table and lateral flowing over it, possibly converging in cones of depression or drawdown regions at wells or streams;
14. Penetration into the aquifer of the soluble, emulsifiable components; and
15. Dispersion and migration within the aquifer and eventual appearance in water supplies of wells, streams, or lakes.

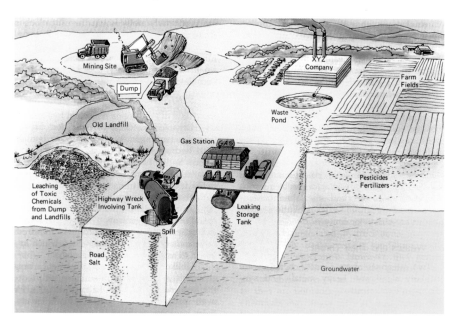

Fig. 15.5. Leakage of chemicals into groundwater (Nebel 1987).

REMEDIATION OF CONTAMINATED SOILS

Contaminated soils can be remediated by leaching, venting or vapor extraction, microbial decomposition, composting, vegetative uptake and removal, and by soil removal. Leaching of soluble contaminants is done by flushing the soil with water and safely draining away the diluted leachate. Microbial decomposition is carried out by organisms in the soil capable of decomposing organic contaminants (e.g., oils), rendering them harmless. The microbial activity can often be stimulated by adding nutrients, aerating the soil if is waterlogged, or irrigating the soil if it is dry.

Composting consists of mixing the contaminated soil with an admixture of readily decomposable organic matter to stimulate microbial activity, then placing the mixture in piles to "cure". The biochemical reactions in the piles typically raise the temperature above 50 degrees C.

Vapor extraction can be done in the case of a soil contamination by volatile organic chemicals such as trichloroethylene from a spilled solvent or benzene from petroleum storage tanks. The soil to be purified is placed on an impervious surface and covered with an impervious cover (generally plastic sheeting). Air is drawn through the soil via perforated pipes, which vents the soil to the atmosphere or to a carbon trap.

Efforts are being made to enlist the soil's own inherent, multifunctional mechanisms in the task of detoxifying contaminants. A common practice in the effort to remediate soils charged with inorganic pollutants is to apply lime to counter soil acidity and thus to suppress the solubility of the contaminants.

Bioremediation is the general name given to a set of practices that enlist plants, microorganisms, or their enzymes to degrade contaminants for the purpose of restoring ecosystem health. Included in this category are techniques of inoculating the soil with desirable cultures and of adding nutrients to spur their growth and activity, specially aimed to degrade a particular type of contaminant. This mode of treatment can be applied to the soil directly or to materials (organic residues and waste products) that are composted or otherwise processed prior to being applied to the soil.

Direct phytoremediation is the use of plants to absorb and remove contaminants from the soil. Some plants tend to concentrate a specific element, such as one or another of the heavy metals, and allow its removal and safe disposal at the time of harvest. Certain species of plants (e.g., the genus Thlaspi of the Cruciferae family, which tends to accumulate zinc, and the genus Phragmites, which grows profusely in a water-saturated soil and accumulates a spectrum of contaminants) have been found to serve the purpose efficiently. Indirect phytoremediation is the process by which the root exudates of certain plants stimulate microbes in the rhizosphere to degrade organic contaminants. It appears that human interventions designed to enhance the complexity and richness of the rhizosphere can help it to fulfill an active role in the restoration of the soil environment.

16. CONCLUDING OVERVIEW

Out of the earth you were taken,
for soil you are,
and unto soil you shall return.
Genesis 3:19

THE SOIL-POPULATION-FOOD DILEMMA

Of the world's land area, only some 25 percent can be considered suitable for agriculture, the remainder being either too steep, rocky, cold, wet, or too dry for the growing of crops. The actual land under cultivation (estimated at some 11.8 percent) is less than half of the potentially cultivable area, with an additional quarter used for grazing livestock, including large areas of prairie, savanna, and scrub vegetation. However, any further expansion of cultivation would pose a severe threat to the remaining natural ecosystems and their vital biodiversity. Hence, to meet the requirement for food security, there is an inescapable need to intensify production and to do so sustainably—that is to say, without degrading the resource base of soil, water, and energy, and without changing the climate for the worse.

At the core of the problem lies the inexorable increase of consumption due to rising populations as well as to the universal desire to attain ever higher living standards. Consequently, our expansive civilization has been placing greater and greater demands on the world's limited and fragile resources.

At the beginning of the eighteenth century, three centuries ago, the world's population totaled less than 600 million. Since then, with infant mortality reduced and life expectancy prolonged, the population has increased tenfold. It had doubled by the middle of the nineteenth century to about 1.25 billion, and then doubled again in the next hundred years. Population growth further accelerated in the second half of the twentieth century, to 5 billion by 1987 and to 6 billion at the beginning of the twenty-first century.

Notwithstanding the recent trend toward reduced fertility in many countries, the momentum of population growth during the last few decades dictates that human numbers will continue to grow, simply because of the increases in the number of young people of fertile age and in the overall life expectancy. World population is now projected to stabilize at some 9 to 10 billion by the third decade of the twenty-first century, but that is only a projection, not a certainty.

Another important demographic change affecting the environment is the changing pattern of human labor and habitation, specifically the growth of urbanization. At the beginning of the twentieth century, the greater portion (some 75 percent) of the world's labor force was employed in agriculture, and the majority of these were subsistence farmers who produced mainly for their own needs. Since that time, as farm work became increasingly mechanized, the portion of the workforce employed on the farm declined greatly. In many of the industrialized countries (including the United States, Europe, and Japan), it has fallen below 2 percent.

The vaunted productivity of modern agriculture has its problematic side, however. Farmers, once largely self-sufficient, now rely on industry for their tools, machines, fertilizers, pesticides (more properly called biocides), electricity, and fuel. The intensive use of these inputs has produced a growing chemical dependency, requiring ever greater doses to compensate for the declines of soil fertility and the adaptive resistance to pests. Accumulating residues of these chemicals have had the effect of contaminating the larger environment, polluting groundwater aquifers and surface water bodies (streams and lakes), decimating wildlife, and threatening the health of domestic animals and humans. Larger machines, operated effortlessly and hence often used carelessly, can cause direct damage to the soil—including compaction and excessive pulverization, leading to accelerated erosion, as well as to damage to aquifers by over-pumping and depletion.

The enormous increase of labor efficiency resulting from the use of motorized machinery has been purchased at the cost of greatly increased consumption of energy and reliance on external, unstable, and increasingly expensive energy sources. In some types of agriculture, the total amount of energy consumed, in fuel needed to operate engines and to produce and supply fertilizers and many other inputs, exceeds the energy value of the products. So while modern mechanized and chemicalized agriculture seems highly efficient from the point of view of human labor, it is remarkably inefficient from the point of view of energy input versus output. Judging by the latter criterion, modern agriculture is, in fact, less efficient than the mode of farming practiced in the unindustrialized countries, which some people—ignoring the energy ratio—consider "primitive".

The growing cost of fossil fuels, and recognition of the negative environmental effects of their excessive use, must induce a change in the mode of agriculture. Much of the energy used in agriculture (for example, in excessive tillage) is in fact wasted and may do more harm than good. The current trend toward minimum tillage, or even zero tillage, is a case in point. Yet much more attention must be devoted henceforth to raising the efficiency of energy use in agriculture, as well as reducing the excessive application of potentially harmful chemicals (fertilizers and biocides).

Current efforts are being made to develop and adopt alternative practices capable of enhancing the ecological resilience and long-term sustainability of farming by relying less on infusions of fuel and chemicals from the outside and more on beneficial natural processes and renewable resources drawn from the farm itself. Such efforts have been described as unconventional, organic, regenerative, or low-input agriculture. They are based on diversifying and rotating crops (polyculture rather than monoculture), building up the soil by enriching it with organic matter, promoting microbial activity that enhances the availability of nutrients and the stability of soil structure, using biological in preference to chemical methods to control pests, minimizing tillage, and utilizing renewable energy (wind power, hydropower, and biofuels).

Agriculture, being the employment of land and water for human needs, is a fundamental aspect of our civilization on which rests not only the quality of human life but indeed its very existence. If the sustainable modes of soil and crop husbandry are further developed and more widely adopted, agriculture should become less intrusive, less disruptive, and more harmonious in the environment. At the same time, land degraded by past abuses must be withdrawn from further exploitation and rehabilitated.

 BOX 16.1 Increasing Food Production

Food production can be increased either by expanding the area under cultivation or by intensifying production on selected lands already under cultivation.

For many centuries, the dominant mode was the first option; i.e., the clearing of more land of its native cover, in the process of which natural ecosystems were destroyed on an ever-increasing scale. This process, carried out at the expense of terrestrial biodiversity, obviously could not be continued indefinitely. Eventually, most of the readily accessible favorable land was appropriated, and much of that land was degraded as a consequence of exploitative and unsustainable modes of management. Henceforth, the main prospect for increasing production is to intensify management of the most favorable land rather than to appropriate ever more marginal and submarginal land.

Such intensification involves a package of measures that must be tailored to the specific circumstances in each case. Those measures may include the use of high-yielding crops and varieties, and a set of treatments designed to optimize growing conditions. The latter include: ensuring adequate water and nutrient supplies (under both rain-fed and irrigated regimes); minimizing losses of water and nutrients (by runoff, percolation, and competition by weeds); providing a favorable rooting zone (that is deep, soft, well aerated and free of toxic and/or pathogenic factors); and protecting the crops against diseases and pests. To be sustainable, however, such advances must be accompanied by measures to protect the soil against degradation, such as might result from soil erosion, waterlogging, nutrient depletion, salination, acidification, pollution, dispersion, compaction, or crusting.

The concentration of population in urban communities and the funneling of products from extensive, sparsely populated hinterlands into densely populated centers have created new problems of transportation, distribution, energy use, domestic and global climate impacts, and waste disposal. As more and more people are moving into cities, the cities are expanding at the expense of some of the best farmland. In the United States of America, for instance, the total area of cropped land has declined from 170 million hectares in 1982 to less than 150 million hectares in 2002. Worldwide, some 20 to 30 million hectares are converted from farmland to urban uses annually at present.

The process of urbanization has its environmental consequences within each city as well as beyond it. Buildings and streets create impervious surfaces that generate a great volume of rainstorm runoff, which in turn carries pollutants, including oil leaked from automobiles and residues of various chemicals used in homes and businesses. Since these pollutants are not filtered through the soil *in situ*, they tend to flow into rivers and reservoirs. Climatically, the preponderance of asphalt, concrete, and glass, and the paucity of actively transpiring vegetation in cities, result in the absorption and emission of heat. Consequently, cities tend to be warmer by several degrees on average than the open countryside, a phenomenon known as the *urban heat island*. At the same time, motorized traffic and industries result in the emission of much smoke and smog-causing gases that affect the radiation balance as well as public health. Wise city planning, therefore, allocates a goodly fraction of the urban expanse to open parks, tree-lined boulevards, and vegetated "green roofs" .

Of the land area, constituting about 30 percent of the globe's surface, only about 11.8 percent is arable. However, that proportion varies from one country to another and from one continent to another. In the Ukraine, about 57 percent of the land is arable. In the US, the fraction is about 19 percent, while in Egypt, it is less than 3 percent. The area per capita devoted to the production of grain (the staple food in many countries) has diminished from 0.2 hectare in 1950 to less than 0.13 hectare in 1990. That decrease—attributable to population growth, land appropriation by expanding urbanization, and soil degradation—is a continuing trend. By the year 2030, the area of grain-producing land per person is expected to average no more than 0.08 hectare. In some countries (e.g., in China), it has already diminished to that extent. These trends and prospects emphasize the imperative to conserve the remaining agricultural land and to improve and intensify production on a sustainable basis without damage to the natural environment.

Although the growth of population undoubtedly contributes to what is now acknowledged to be an environmental crisis, an even more important factor is the nature of human activity. Some of the most crowded countries in the world (such as the Netherlands and Japan) have managed their environments with greater care, while others (such as the United States) have not done so well. Profligate energy use in industry, in transportation, and in domestic life, as well as carelessness in the use of materials and in the disposal of wastes, characterize the hasty economic "progress" of rapidly industrializing countries such as China and India, as well as the US and Russia. The result is pollution of air, soil, and freshwater resources, and even of seas. Not the least of the consequences, now recognized as global, is the threat of climate change.

The transformation of the planetary environment induced by this explosion of human activity is particularly evident in the changes wrought upon the landscape. Since the beginning of the eighteenth century, the planet has lost six million square kilometers of forest. Land degradation has increased significantly, as have the sediment loads of river systems. During the same period, the amount of water withdrawn by humans from the natural hydrological cycle has increased from about 100 to some 4500 cubic kilometers per year. In the last two centuries, agricultural and industrial development have nearly trebled the amount of methane in the atmosphere and increased the concentration of carbon dioxide by some 40 percent. Issues of economic growth versus ecological preservation that formerly were considered locally autonomous are now seen to involve multiple global linkages among such activities as energy consumption, water use, industry, transportation, urban versus rural development, agriculture, forest clearing, drainage of wetlands, management of floodplains and coasts, as well as waste disposal and control of climate change.

The dilemma of striving to satisfy an increasing demand in the face of limited—and in some cases dwindling—resources has worried many observers ever since the Reverend Thomas Robert Malthus wrote his *Essay on the Principle of Population* in 1798. In it, he argued that population tends to increase faster than food supply, and that, unless that increase is checked by moral restraint, it must inevitably lead to more wars, famine, and disease. For a time, however, the Malthusian warnings seemed exaggerated, even wrong. In the last three or four decades of the twentieth century, food production grew faster than population in all the continents except Africa. That increase, dubbed the "Green Revolution", resulted from the close cooperation between plant breeders and soil scientists. The plant breeders developed new varieties of grain crops (primarily wheat and rice) with higher yielding potential, which depended, however, on optimal conditions of soil moisture and nutrients. The latter were the contributions of soil scientists, whose research resulted in improved soil management practices based on the optimal applications of fertilizers, soil amendments, and irrigation.

The great improvements of the last few decades, however, may not continue indefinitely. Although new methods of genetic engineering, based on recombinant DNA and other innovations in biotechnology, offer great promise (albeit not without attendant risks), there are constraints to expanded production resulting from the fact that the most favorable soils and readily accessible water resources (i.e., easily tapped aquifers and favorable sites for damming rivers) have already been appropriated. Further expansion of the area under cultivation and further diversion of water flows can only be achieved at the cost of disrupting the remaining natural ecosystems. Much of the land yet unutilized is marginal in terms of its potential and vulnerable to degradation by such processes as erosion, compaction, organic matter and nutrient depletion, pollution, salination, desertification, and loss of soil biodiversity.

It must be emphasized and reemphasized that soils are inherently fragile. What takes nature hundreds or thousands of years to create and enrich can be degraded or destroyed in just a few years by careless or short-sighted human mismanagement. Hence, the importance of soil conservation is paramount, and is, in essence, in the interest of humanity's self-preservation.

Another constraint to increasing production is the rising cost of fuel. Mechanized agriculture is extremely energy-consumptive. Its great increase in output per man-hour of labor has been achieved at the expense of much greater reliance on fuel-driven mechanization. When the cost-benefit ratio is calculated in terms of energy input and output, the balance of modern agriculture is, in many cases, negative. With the rising cost of energy, that mode of agriculture is clearly unsustainable economically. Worse yet, the profligate use of fossil fuels has resulted in a significant and progressive increase in the concentration of radiatively active trace gases in the atmosphere, an effect likely to increase the temperature and very possibly the frequency and severity of weather anomalies—storms and droughts—that will diminish rather than enhance food security. Global warming is also likely to cause sea-level rise, which may well cause the waterlogging and inundation of currently productive and habitable coastal lands.

One mark of our society's profligacy is the luxury consumption of animal products. Given that some 10 kilograms of grain are needed to produce just 1 kilogram of meat, the dietary habits of the rich countries (which, incidentally, are not necessarily healthy) seems extremely wasteful in terms of the soil, water and energy resources needed to satisfy them.

The one continent that has not yet taken part in the Green Revolution is Africa. In 1967, the annual grain production there averaged about 180 kilograms per person, an amount regarded as barely adequate to sustain healthy life. Two decades later, instead of rising, the average production had fallen to 120 kg per person. That unfortunate situation is both a cause and a consequence of strife in a continent that is the original home of the human species.

The problems of Africa are complex. Much of the continent is arid or semi-arid, with fragile soils that are extremely vulnerable to drought and to erosion. In the parts of the continent that are relatively humid—i.e., the central tropics—the soils are highly weathered, leached of nutrients, and affected by aluminum toxicity. Over large areas, the deep-rooted vegetation that had recycled nutrients from the lower layers of the soil profile and that had returned them to the surface zone was cut and used for construction or was burned for heating and cooking. Thus, the soil was deprived of its protective and self-restorative cover of deeply rooted vegetation, surface mulch, and organic matter. As the original organic matter decomposed and nutrients were extracted without replenishment, the soils lost their original fertility and their structure was degraded.

An effective method to restore soil fertility is the practice known as agroforestry, by which crops and trees are grown together. Fast-growing nitrogen-fixing (leguminous) trees can be planted in parallel rows, preferably on the contour, with crops grown in between. The trees act as nutrient pumps, improving soil fertility and providing forage as well as firewood, while the crops provide food for subsistence or local marketing. In some cases, the tree rows are cut down every few years for timber and fuel, and their strips rotated with the alternating cropping strips.

Much can be done in Africa (as elsewhere) to improve water conservation and water-use efficiency, both in rain-fed farming and in irrigated farming. Modern methods of low-volume, high-frequency irrigation can be applied to small-holder farms, using inexpensive, locally fabricated equipment. Investments are needed, however, to develop the infrastructure of water supplies and distribution.

BOX 16.2 Soil Degradation: A Summary

So complex and intricate are the functions of the soil that there are multiple ways, often interrelated, for them to be disrupted and degraded.

One process is the disruption of soil structure: the slaking and breakdown of aggregates and dispersion of the clay fraction, leading to crust formation and consequent inhibition of water entry, aeration, germination, and seedling establishment.

Often closely related is the mechanical disruption of the soil due to trampling by animals or—more often—traffic by heavy machinery. When the soil is moist, such traffic results in compaction, and when the soil is dry traffic results in pulverization of aggregates, leaving the surface vulnerable to deflation by wind and scouring by raindrops and running water.

Another process is the depletion without restoration of the soil's reserve of organic matter by decomposition (oxidation in aerated soils or reduction in waterlogged soils), which also entails loss of nutrients.

Soluble nutrients are often lost by leaching under conditions of excessive percolation. Such conditions may also restrict aeration.

The opposite of excessive leaching is the excessive accumulation of soluble compounds in the soil, especially of toxic factors such as heavy metals and pesticide residues.

The burning of vegetation and its residues often causes the extreme dehydration of the soil surface, which may thereby become, at least temporarily, water-repellent. Consequently, water infiltration may be reduced, runoff increased, and erosion may result.

Erosion is perhaps the most widespread and damaging mode of soil degradation. It results in the loss of soil fertility, particularly as it deprives the soil of its most fertile layer, the A-horizon (known as the topsoil).

Still other modes of soil degradation may involve waterlogging, acidification, salination, sodification, and the accumulation of toxic elements and compounds.

Of all soil layers, the surface zone is the most critical. It is the avenue through which water and nutrients enter, runoff is generated, gases are exchanged, and seeds germinate. It is also the zone where organic matter accumulates by the decomposition of plant and animal residues, or is depleted by decomposition. Finally, it is the zone most vulnerable to compaction and erosion.

PROTECTION OF SOIL RESOURCES

Good soil must be regarded as a limited and precious resource. It must provide for human needs even while continuing to serve as the habitat and the source of sustenance for the entire community of terrestrial life. The fact that humankind, originally a child of the natural environment, has been able to gain mastery over the habitat, has in the final analysis made all life on earth more precarious rather than more secure.

There is an old adage concerning the difference between the clever and the wise. The clever are those who are able to extricate themselves from situations that the wise would have recognized and avoided from the outset. Now, however, it seems that our short-sighted cleverness as a species has gotten us into a quandary that mere cleverness can no longer resolve. Wisdom is now needed more than ever before.

The wisdom we need can only be developed through interdisciplinary research and active international cooperation in the protection of our shared environment. The ultimate purpose of environmental activity should be to ensure that each generation bequeaths to its successors a world that has the full range of natural wealth (enhanced, insofar as possible) and the richness of human potentialities that it had received from its predecessors. That range encompasses the land, its soils, its waters, its atmosphere, its energy resources, its raw materials, as well as its multifarious and synergistic forms of life.

Present yields in many areas are much below the proven potential yields. Where they can be enhanced substantially, there should be no need to claim new land and to encroach further upon natural habitats and their biodiversity. The possibilities for developing intensive, efficient, and sustainable agriculture can obviate the need for the widespread cultivation and grazing of marginal lands, thus allowing the regeneration of natural habitats.

A major challenge of our time is to achieve harmony between the responsibilities and needs of developed and developing nations, between the needs of our generation and those of future generations, and between the human species as a whole and other species in the community of life on earth. Ecology teaches us that each member of a community is defined not by its individual traits alone but by the nature of its reciprocal relationships with other sharers of the same domain. The ancient tribal vision of the world is still deeply ingrained in us. Gradually, however, our vision has evolved and our notion of kinship has extended to include first our immediate clan or tribe, and then successively our village, city, country, nation or creed, and—eventually—all of humanity. This expanded perception of kinship and allegiance must now transcend the bounds of the human species and extend to the totality of life on planet earth.

We cannot continue or even tolerate practices that cause erosion, salination, groundwater depletion and contamination, or policies that keep poor nations permanently dependent on rich nations. We must stop the destruction of habitats, the eradication of species, and the disruption of biotic communities.

SOILS IN THE ENVIRONMENT: CONCLUDING STATEMENT

Soils perform multiple functions of water and nutrient cycling and of gas exchange with the atmosphere. They support all forms of terrestrial vegetation (trees, shrubs, grasses, algae, etc.), which, in turn, support animal life. As more and more land has been brought under human control for cultivation, grazing, or construction (roads, houses, factories, seaports, airports, and recreational amenities), the land remaining for natural ecosystems has been diminished, fragmented, and impoverished. The task before us now is to preserve and enhance the remaining natural ecosystems by restricting the further usurpation of land by human endeavors. Wherever possible, we must rectify the damage done by restoration of natural habitats and by relieving pressure on marginal or degraded lands. To do so, we must intensify and improve the efficiency of production on the best managed lands, and do so in ways that are sustainable, without damaging the adjacent environmental domains. For example, fertilizers are to be used in measured doses and at rates designed to answer the variable needs of crops and to promote soil fertility while avoiding wasteful leaching and runoff that may contaminate aquifers, streams, and lakes.

The protection and nurturing of the soil must be recognized as a principal component of environmental stewardship. Augmenting the organic matter content of soils can contribute not only to mitigating climate change but also to the improvement of agricultural productivity and environmental quality.

In a larger context, the soil is not merely a passive medium, privately owned and operated by farmers to produce crops, but is indeed a common environmental asset, providing ecosystem services such as support for wildlife and transformation of pollutants. As such, the soil must be protected by purposeful policy to prevent its erosion, loss of organic matter, compaction and sealing, and contamination. This should apply not only to the agricultural sector but also to soil in the urban sector that is especially endangered by pollution. Positive inducements are necessary, along with mandatory regulations, to ensure soil's ecological and functional integrity.

Soil is a finite, fragile resource. Its proper treatment is a condition for the survival and success of human civilization. In the words of Franklin Delano Roosevelt (1934): "A nation that destroys its soil destroys itself." We choose to paraphrase his statement in positive terms: *A nation that saves its soils saves itself.*

In the words of Genesis 2:15: "The Lord God placed the human Earthling (Adam) in the Garden of Delight (Eden) *to serve and preserve it.*"

BOX 16.3 Summary of Soil Functions in the Environment

Regulation of the Hydrological Cycle:
    Partitioning of rainfall
    Infiltration vs. surface-water excess
  Initiation of surface runoff
Routing of surface runoff
    Distributed overland flow
    Concentrated flow in rills and gullies
Storage and disposition of water in the root zone
    Direct evaporation from the surface
    Uptake by roots, transpiration by plants
Drainage beyond the root zone
    Fluctuation of water-table level
    Recharge of subterranean aquifers
    Discharge via springs, rivers, and wells
Cycling of Soluble Components:
    Hydration, dissolution, re-precipitation of minerals
    Sorption and exchange of ions
    Decomposition, re-composition of organic compounds
    Oxidation-reduction reactions
    Acidification-alkalization reactions

Salination-desalination processes
Eluviation-illuviation of solutes in the soil profile
Leaching of solutes to groundwater and streams
Volatilization and outgassing to the atmosphere
Cycling of Particulates:
Filtration of suspended particulates in percolation
Migration and lodging within the soil profile
Clay-coating of aggregates
Accretion (deposition) of air-borne particles (aerosols)
Erosion of the soil surface by water and wind
Overland transport of water-suspended matter
Deposition of suspended matter in depressions/estuaries
    Silting of lakes and reservoirs
Cycling of Energy:
Absorption of incoming solar (shortwave) radiation
Reflection of a fraction of solar radiation (albedo)
Emission of terrestrial (longwave, thermal) radiation
Transmission and exchange of sensible heat
    Biota, atmosphere, hydrosphere
Mitigation or enhancement of the greenhouse effect
    Assimilation of chemical energy in biomass
    Emissions of CO_2, CH_4, N_2O (greenhouse gases)
Sustaining Biota:
Providing substrate, water, nutrients to microbial community
Providing anchorage, water, nutrients to roots of higher plants
Regulating biotic respiration (O_2, CO_2 exchange)
    Aerobiosis and anaerobiosis
Decomposing plant and animal residues, release of nutrients
Absorbing and neutralizing pathogenic and toxic agents

APPENDIX A: THE ROLE OF SOIL IN THE MITIGATION OF GLOBAL WARMING

BACKGROUND

The earth receives a flux of short-wave radiation from the sun. It reflects a fraction of that radiation and absorbs the greater part of it. The absorbed radiation tends to heat the surface of the earth, which radiates an outgoing flux of long-wave radiation to outer space. The balance of incoming and outgoing radiant-energy fluxes determines the temperature of the earth's surface. If the atmosphere blanketing the earth were transparent to the outgoing long-wave radiation, retarding none of it, the equilibrium mean surface temperature of the earth's surface would be a frigid −18 degrees Celsius. In reality, however, the outgoing radiation is retarded by water vapour (which makes up about 1% of the atmosphere on average), by cloud droplets, and by several trace gases (so-called because they are present in small, or "trace" concentrations). That atmospheric effect raises the mean surface temperature of the earth to about 15 degrees Celsius. The absorptive trace gases that occur naturally in the atmosphere are water vapour (H_2O), carbon dioxide (CO_2), ozone (O_3), methane (CH_4), and nitrous oxide (N_2O).[1] Together, these radiatively active gases are known as the atmospheric "greenhouse gases" (GHGs).

---

[1] Another group of radiatively active gases introduced artificially into the atmosphere are the synthetic chlorofluorocarbons (CFCs), the further manufacture of which has been banned by international agreement (the so-called Montreal Protocol) since 1987.

The natural greenhouse effect, such as it is, makes the world more hospitable to life forms than it would be otherwise. However, the progressively rising concentrations of some of these gases poses the danger of excessive global warming. That rise is due mainly to combustion of fossil fuels (coal and petroleum) to clearing of natural vegetation and to cultivation and erosion of soils. Three of the GHGs (namely, CO_2, CH_4, and N_2O) are especially involved (to the extent of >80%) in the dramatic increase of the atmosphere's greenhouse effect, evidently due to human action, which has taken place increasingly ever since the start of the Industrial Revolution. Most of the discussion to date has focused on the role of fossil fuels and their combustion products as they might affect the large-scale atmospheric and oceanic phenomena that control the world's weather. Less attention has been devoted to the processes taking place within the terrestrial domain, which is inherently complex and heterogeneous, given its array of soils, vegetation, and modes of management, as it affects the dynamic balance of GHG emissions to, or—conversely—absorption from, the atmosphere.

The most abundant and hence the most important of the human-affected greenhouse gases is carbon dioxide. Its concentration in the atmosphere has increased in the last two centuries from about 275 to some 385 parts per million on a volume basis (ppmv), and it continues to increase at the average rate of well over 1 ppmv per year. CO_2 is a biologically essential gas, as it is absorbed from the atmosphere by autotrophic green plants and combined with soil-derived water in the all-important process of photosynthesis. That process, which is powered by radiant energy from the sun, produces the basic chemical constituents of life (carbohydrates, oils, and proteins). Innumerable heterotrophic forms of life on earth derive their energy from the decomposition of those products in a process called *respiration*, which is the biochemical reversal of photosynthesis.

The second soil-emitted greenhouse gas is methane, CH_4. It is present in the atmosphere at a much lower concentration than CO_2, yet its specific effect (molecule-per-molecule) is some 20-fold stronger. At present, it accounts for about 16% of the warming effect due to the rising concentrations of greenhouse gases. In the last two centuries, its atmospheric concentration has more than doubled, and is currently increasing by about 0.3% annually. Its main sources are wetlands (both natural and managed, the latter mainly consisting of flooded rice paddies), the enteric fermentation of ruminant livestock, and manure storage in saturated lagoons. In well-aerated soils, methane oxidizes to form CO_2.

The third pedogenic greenhouse gas is nitrous oxide, N_2O. It occurs at an even lower concentration than methane (about 1/1000 the concentration of CO_2), but its specific (molecule-per-molecule) warming effect is about 300 times greater. Its concentration has been increasing at the average rate of about 0.2–0.3% annually, and about 40% of this increase is attributed to agriculture—mainly the excessive or untimely application of nitrogenous fertilizers, especially in poorly aerated soils.

At present, there seems to be no realistic way to return the atmosphere's GHG concentrations to their pre-industrial levels within the next few decades. Even to stabilize those concentrations at twice the pre-industrial level (i.e., at a level equivalent to 550 ppmv of CO2) will require a two-thirds reduction of GHG emissions from the projected level of 21 billion tons per year to 7 Gt (measured as C) by, say, mid-century. Such a task will require greater energy-use efficiency, reliance on non-carbon energy sources (such as solar, wind, hydroelectric, and nuclear), and efforts to absorb and sequester carbon from the atmosphere in geologic formations, the deep ocean, and the soil.

The present storage of organic carbon in the world's soils has been estimated to total between 1500 billion metric tons, or Gt (to 1-meter depth) and 2400 Gt (to 2-meter depth). This is compared to the atmospheric content of some 750 Gt, the terrestrial-biotic (mostly vegetative) content of some 560 Gt, and the much larger oceanic content of some 5000 Gt. Note that the total amount of carbon in the soil is over four times as large as the amount in terrestrial biota and over three times as large as the amount in the atmosphere. In addition to the organic carbon, the soil also contains large amounts of inorganic carbon (SIC), mostly in the form of calcium and magnesium carbonate, estimated to total some 695 to 748 billion tons, present mainly in the soils of semiarid and arid areas. Though not nearly as labile as organic carbon, SIC can be solubilized by acid and is subject to leaching. Some carbon dioxide also dissolves in groundwater, and may be released to the atmosphere by effervescence as, for example, when groundwater is pumped up and used for irrigation.

Many soils around the world are already overexploited and degraded, so that their capacity to support diverse forms of life is partially impaired. Global warming will entail a series of interactions involving soils that may cause their further deterioration. Higher temperatures tend to induce higher rates of organic matter decomposition, which not only deprive the soil of nutrients and impair its physical functioning but also result in greater emissions of carbon dioxide into the atmosphere that, in turn, further exacerbate the tendency toward soil warming. Climate change is also expected to intensify weather phenomena such as droughts that may denude the soil of its protective vegetative cover and may alternate with violent rainstorms that will erode the soil.

As mentioned above, the world's soils and the vegetation they support contain an enormous quantity of carbon, far more than is contained in the atmosphere in the forms of the greenhouse gases CO_2 and CH_4. The soil's store of carbon is labile rather than permanent; it can be augmented by net absorption from the atmosphere, or reduced by net emissions to the atmosphere due to decomposition. Carbon losses from the soil may also be due to erosion of the topsoil, as well as to leaching into the subsoil or groundwater. The balance of soil carbon is greatly influenced by anthropogenic factors, including the clearing or restoration of natural vegetation and the patterns of land use in the agricultural, industrial, and urban areas. Cultivation, especially, spurs microbial respiration and decomposition of organic matter. In aerated soils, organic matter oxidizes and releases CO2; in poorly aerated soils, SOM undergoes chemical reduction and tends to release CH4 (as well as N2O). Consequently, some soils may eventually lose as much as one-third to two-thirds of their original organic-matter content.

Though agricultural soils acted in the past as significant sources of atmospheric CO_2 enrichment, they offer an opportunity at present and in the coming decades to effect the substantial absorption of CO_2 from the atmosphere and its storage as added soil organic matter. The historical loss of carbon in the world's agricultural soils has been variously estimated to total some 42 to 78 Gt. In an ideal world, we might hope for complete restoration of that loss; i.e., a return to a pristine state of "carbon saturation". In reality, soil degradation resulting from tillage, erosion, and other processes has diminished the capacity of soils in many areas to fully recover their original state in the period of time of our consideration (say, the next few decades). Even where such restoration is possible, its attainment may not be economically feasible. The actual carbon-sink capacity of many soils (i.e., the potential restoration of their carbon content in practice), assuming the adoption of recommended technologies and strategies of soil management, may be of the order of half to two-thirds of the historic C loss. Still, that amount is significant.

BOX A.1   Fire Setting as a Traditional Mode of Land Management

Extensive areas of pastoral grassland and cultivated land (e.g., in the Americas, Australia, Africa, and Southeast Asia) have traditionally been maintained by the purposeful setting of periodic fires to suppress weeds and to prevent the encroachment of woody plants. In such savanna ecosystems, the balance between trees and grasses, stand structure and dynamics, and shrub cover and abundance is determined to a large extent by the frequency and intensity of fire.

Productivity often increases temporarily following fire as a result of microclimatic modification (due to removal of litter and standing crop) and greater nutrient availability. However, such fires deprive the soil of the organic matter that could otherwise protect and enrich the soil. The fires not only cause the episodic release of carbon dioxide but also tend to increase soil temperature (due to the elimination of shading by vegetation and litter) and hence to increase the rate of decomposition of soil organic matter. Moreover, by desiccating and baring the soil even temporarily, fires may also induce crusting and accelerated soil erosion. A compensating factor may be that the nitrogen-fixing function of certain leguminous plants (e.g., mesquite) may be stimulated by fires.

The potential sequestration of carbon in global agricultural soils through changes in management practices has been variously estimated to total between 600 and 900 Mt C per year over a period of several decades. The recommended practices include reforestation, agro-forestry, no-till farming, cover crops, nutrient augmentation (with fertilizers, manures, composts, and sludge), soil amendments (e.g., lime to neutralize excessive acidity), improved grazing, water conservation, and the production of energy crops to replace fossil fuels. If implemented on a large scale, such practices may help to mitigate the greenhouse effect, boost crop yields, reduce soil erosion, improve soil structure and water quality, enhance biodiversity, and help to provide food security. However, these potentialities vary greatly from one soil type and location to another and depend on such properties as soil texture, soil profile characteristics, soil-water relations, soil aeration, vegetation density and type, and the prevailing climate.

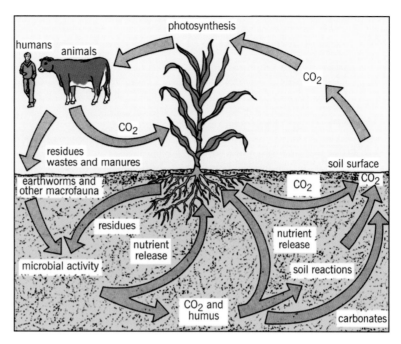

Fig. A.1. The carbon cycle in the soil (schematic). (Dubbin, 2001).

Table A.1 Estimated mass of carbon in the world's soils (excluding glacier-covered areas)

Soil orders	Area k.km2	Area %	Organic C t/ha	Organic C global Gt	Organic C % global	Inorganic C Gt
Alfisols	13,159	10.1	69	90.8	5.3	43
Andisols	975	0.8	306	29.8	1.8	0
Aridisols	15,464	11.8	35	54.1	3.2	456
Entisols	23,432	17.9	99	232.0	13.7	263
Gelisols	11,869	9.1	200	237.5	14.0	10
Histosols	1,526	1.2	2,045	312.1	18.4	0
Inceptisols	19,854	15.2	163	323.6	19.0	34
Mollisols	9,161	7.0	131	120.0	7.0	116
Oxisols	9,811	7.5	101	99.1	5.8	0
Spodosols	4,596	3.5	146	67.1	3.9	0
Ultisols	10,550	8.1	93	98.1	5.8	0
Vertisols	3,160	2.4	58	18.3	1.1	21
Other soils	7,110	5.4	24	17.1	1.0	5
TOTALS	130,667	100.0		1,699.6	100.0	948

Note: five soil orders (Entisols, Gelisols, Histosols, Inceptisols, and Mollisols) account for some 72% of all the organic carbon in the world's soils. Gelisols alone account for between 14% and 24.5% of the total. There is, however, a large measure of uncertainty in the data.

BOX A.2   A Historical Perspective

Control of the atmospheric greenhouse effect and of global warming requires a thorough understanding of the complex exchange processes involved in the natural carbon cycle and the ways it has been influenced by human activity over time. Prior to the middle of the twentieth century, the release of CO_2 into the atmosphere from soils and vegetation had apparently exceeded the release from the burning of fossil fuels. At present, the contributions to the atmosphere of CO_2 from deforestation and land use are estimated to total 1.5 billion tons per year, which constitute over 20% of the total anthropogenic emissions of about 7 billion tons per year. The net emissions from soils alone are likely to be in the order of 0.5 billion tons per year. Fossil fuel burning currently accounts for about 50% of the total CO_2 release to the atmosphere, with biomass respiration, decay, and burning emitting the remaining 50%. Carbon sinks such as terrestrial plants and ocean biota take up most of the emitted CO_2.

The concentration of CO_2 in the atmosphere has increased by over 35% in the last two centuries, from about 275 ppmv to some 385. It is now increasing at an average rate of about 1.5 ppmv (~ 0.4%) per year. Atmospheric concentrations of CH_4 and N_2O have been increasing progressively as well. The radiative forcing of CO_2 is estimated to total $1.46 W/m^2$, accounting for about 60% of the total greenhouse-gas forcing. The corresponding values for CH_4 are $0.48 W/m^2$ and 20%, and those for N_2O are $0.15 W/m^2$, constituting some 6% of the total force.

It seems unrealistic to expect that the further rise of CO_2 concentration can be halted entirely and immediately. Rather, a realistic goal might be to slow the rate of rise progressively and to stabilize it at about 450 ppmv by mid-century. Even achieving this modest goal will require application of a wide range of technologies and modifications of existing practices, including modes of environmental (including soil) management.

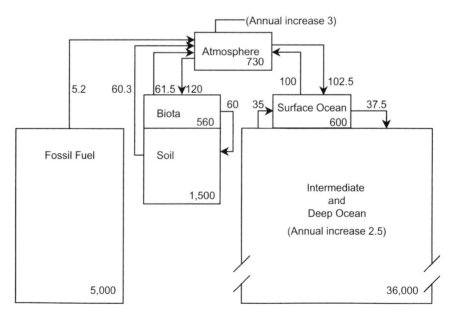

Fig. A.2.   The major reservoirs and flows of carbon in the globe (From Rosenzweig and Hillel, 1998).

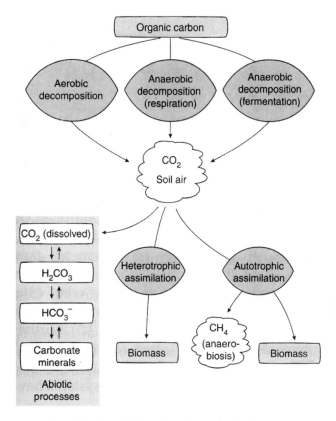

Fig. A.3. Carbon dioxide budget of the soil (Gobat et al., 2004).

AGRICULTURE'S ROLE IN GLOBAL WARMING

The terrestrial domain covers some 30% of the globe's surface. Soils cover roughly 85% of the terrestrial domain, and about 37% of the soil area is subject to agricultural use as cropland or pasture. Agricultural land occupies some 5020 Mha (FAOSTAT, 2006), of which 69% is in pasture and about 28% under cultivation. Those areas have expanded during the last four decades by the conversion of forest land and of other land to agriculture at the rate of about 13 Mha per year. The extent of cropland worldwide is currently estimated to total about 1400 Mha, having increased by 8% in the last four decades. Agriculture contributes some 24% of global GDP (World Bank, 2003) and employs as many as 1.3 billion people. Cropland has expanded by 8% since 1960 – mainly in the developing countries of South America, Africa, and Asia. By some predictions, the area of crops may increase by an additional 300-500 Mha (via continued conversion of natural ecosystems) by the year 2020, mainly in Latin America and Sub-Saharan Africa. Even so, considering the expectable growth of population, the area of arable land per capita is destined to diminish considerably.

Agricultural lands include lands under annual or perennial crops, managed rangeland (pastures), lands devoted to agro-forestry, and – lately – lands dedicated to the production of bio-energy. Historically, the initial conversion of land under natural vegetation to agricultural land resulted in large emissions of greenhouse gases, mostly CO_2, due to the progressive decomposition of the native soil's organic matter. The net emission of CO_2 has diminished in recent decades, and in some areas has been reversed, as reforestation and conservation farming have brought about the partial restoration of soil carbon stocks. At present, the annual emission of CO_2 from agriculture is estimated to total some 40 Mt, constituting less than 1% of the global anthropogenic CO_2 emissions (due mainly to fossil fuel burning). However, agriculture accounts for a sizable share of non-CO_2 emissions, including an estimated 47% of CH_4 and as much as 84% of N_2O.

Taking a positive view, we may surmise therefore that agricultural soils present a significant opportunity for greenhouse gas mitigation through reductions of emissions, as well as through C sequestration. This can be done by promoting the increased absorption of CO_2 by green plants and its stable storage in the soil, as well as by reducing the emissions of the other GHGs. Both objectives can be achieved by improving agronomic practices.

In principle, soil organic carbon (SOC) content is a function of the balance between the rate of organic matter input to the soil (due to the net primary productivity of active vegetation) and the rate of organic matter decay. The rates of these processes differ in space and time, as well as in their sensitivities to varying temperature and moisture regimes resulting from management and climate changes. Mature natural soils contain stocks of organic matter accumulated over many centuries. In such soils, the organic matter had attained a more-or-less stable balance with the rate of its decomposition by soil-borne heterotrophic organisms. After such soils are cleared for agriculture, their cultivation accelerates decomposition of organic matter, and hence the release of GHGs. Moreover, crop removal causes additional loss of biomass that would have otherwise remained and augmented soil organic matter. Erosion, which removes topsoil, further deprives soils of their organic matter. Consequently, the soils degrade in quality, as their fertility diminishes and their structure is destabilized. Agricultural systems that are significant GHG sources, being highly managed systems, are therefore significant potential targets for mitigating the atmospheric greenhouse effect.

A prime example of agricultural practices affecting greenhouse-gas emissions is the production of rice. It is a major crop in Asia, planted on some 130 million hectares, most of which are in flooded paddy. Rice paddies account for about 11% of total global methane emissions to the atmosphere. China has approximately 20% of the world's rice paddies and grows about 31% of the world's rice crop. Over the last two decades, midseason drainage as a modified water management practice has been adopted throughout China. This practice, in contrast to continuous flooding, has been found to raise rice yields by increasing N mineralization and root development, while improving water-use efficiency. It also appears to reduce CH_4 emissions (helping to reduce CH_4 emitted from China's rice paddies by about 40%, or about 5 Tg per year) but to increase the amount of nitrous oxides emitted. The latter emission, induced by midseason drainage, could offset about 40% of the benefit gained by CH_4 reduction. Of the various nitrogenous fertilizers, ammonium sulfate appears to emit the least amount of N_2O.

Any change in management practices (involving water, fertilization, tillage, etc.) inevitably alters the soil environment affecting crop growth and yield, as well as bio-physico-chemical processes (organic matter decomposition, nitrification or denitrification, fermentation, etc.) pertaining to gaseous exchanges with the atmosphere. So complex are within-soil processes and their interactions with the plants and atmosphere that the results of any particular change in management practices cannot be predicted in general but depend on specific conditions. In particular, the adoption of no-till farming entails the problem of pest (primarily weed) control, which – instead of being done mechanically by tillage – is achieved by means of chemical herbicides and other types of pesticides. Regular use of such biocides not only consumes energy and involves the emission of GHGs (directly or indirectly), but may also harm biodiversity.

Among the practices most amenable to improvement are nutrient application and utilization, tillage and soil-structure management, irrigation and water-use efficiency, management of crop residues and of animal manures, control of pests and diseases, breeding of crops and varieties, modification of microclimates, restoration of degraded lands, use of cover crops and agroforestry, and the management of wetlands (including peaty marshlands and flooded rice paddies). In addition, agriculture can help mitigate the greenhouse effect by producing biofuels (from crop residues, manure, and dedicated energy crops) capable of offsetting the excessive reliance on fossil fuels.

Various possibilities exist for creating positive social and economic incentives for mitigation of the greenhouse effect in the practice of agriculture and other modes of land management. First and foremost among such possibilities is a program of extension and education, designed to inform the public at every level.

The task ahead requires an increase in productivity, prevention of erosion and other modes of land degradation, reduction of fuel use in agriculture and of greenhouse emissions, prevention of biomass burning, optimization (spatial, temporal, economic) of the application of fertilizers and irrigation, minimization of tillage operations, augmentation of soil organic matter (including the introduction of species with higher productivity and greater deposition of organic matter within the soil), and conversion of marginal land to agroforestry or the complete restoration of its native forest cover.

Special care is needed in the management of organic soils. Such soils typically form when flooding results in a deficiency of oxygen, which in turn promotes the accumulation of incompletely decomposed organic matter and the emission of CH_4. When converted to agricultural use, these soils are generally drained, and the consequent aeration accelerates decomposition and enhances the emissions of CO_2 and often of N_2O as well. Careful management of soil-water relations and the maintenance of a shallow water table can often reduce those undesirable emissions.

An important principle to emphasize is that climate, soil, and economic conditions differ greatly from one location to another and from one period to another. Therefore, there can be no universal prescriptions regarding practices to manage soils so as to help mitigate the greenhouse effect. While the basic principles can be stated in universal terms, their application to different sites will require specific adjustments. Over time, practices designed to augment soil carbon content are likely to diminish in efficacy, as the soil approaches an equilibrium state or as its organic carbon content attains effective saturation under a particular climate and land-cover. (In fact, there is even danger that the gains of soil-carbon achieved over years or decades may be reversed by reverting even temporarily to older tillage methods or by an outbreak of fire). However, other advantages to C-conservation management, such as reduced energy use and the production of biofuels as substitutes for fossil fuels, can continue indefinitely.

BOX A.3   The Possible Role of Inorganic Carbon in Soils

In addition to organic carbon, many soils contain large amounts of inorganic carbon. Soils in semiarid and arid regions often contain the minerals calcite and dolomite, consisting of calcium and magnesium carbonates. Some of the carbon in such soils is more stable (less labile) than is organic carbon, but may solubilize to some degree and even volatilize (releasing carbon dioxide) when acidified. Where rainwater is particularly acidic (as when it dissolves atmospheric sulfur and nitrogen oxides in addition to carbon dioxide), it may dissolve mineral carbonates in the soil and release CO_2. Under different conditions, organic carbon may precipitate in the presence of calcium to form mineral carbonate. There is, however, little quantitative information on the possible role of soil-carbonate minerals in the soil's CO_2 exchange with the atmosphere.

PAST LOSSES AND POTENTIAL RESTORATION OF CARBON IN TERRESTRIAL ECOSYSTEMS

The combined losses from the earth's native biomass and soils due to deforestation and cultivation during the past three centuries (1700 to 2000) has been estimated to total about 170 Gt of carbon, much of which has been absorbed in the ocean and some of which has accumulated in the atmosphere. Continuing land clearing for agriculture in the tropics apparently results in additional emissions in the order of 1.6 Gt of carbon per year. A shift from deforestation to reforestation began in Europe nearly two hundred years ago and in the northeast US nearly a hundred years ago. In other parts of the world, however, deforestation is expanding, with serious consequences to biodiversity, soil erosion, and global warming.

Global soil inventories allow estimation of the amounts of carbon stored in soils before their cultivation was begun. The 1.7 billion hectares converted to agriculture originally contained an estimated 210 Gt of organic carbon. Assuming an average loss of some 26% from the top 30 cm of mineral soils due to their cultivation, the estimated historical loss of soil carbon may have amounted to some 54 Gt. Substantial (though not total) recovery of that loss could be possible if the degraded lands were returned to their natural vegetation. In practice, only a partial recovery (say, 2/3 of the original carbon content) may be attainable during the next few decades. Moreover, the added carbon will always be vulnerable to secondary or tertiary loss following its unstable sequestration. Especially difficult to restore are the depleted carbon reserves in exploited tropical soils, owing to the high rates of decomposition and erosion in the tropics.

The important principle is that improved management of soil organic matter is a worthy task in itself, beyond its direct benefits in mitigating the atmospheric greenhouse effect. So is the potentiality for the sustainable production of biofuels, especially in view of the imperative to reduce the combustion of fossil fuels, the supplies of which are inherently unsustainable. Even if it appears that agricultural management (or land management in general) can only contribute marginally to mitigation of the atmospheric greenhouse effect, turning the soil from a net source to a net sink for greenhouse gases is still worthwhile. Since there is no single activity that will solve the dilemma of global warming, every sector of human activity must contribute its share to the overall task.

Various feedback mechanisms are involved in the interactions between climate change and the carbon cycle. Increasing concentrations of CO_2 in the atmosphere can stimulate greater rates of photosynthesis, an effect called CO_2 *fertilization*. The carbon thus fixed is allocated partly to labile biomass that recycles readily (e.g., foliage) and partly to more stable cellulose (as in wood). With all plants, some organic carbon is transferred via surface litter and the root system to the soil, and a fraction of that is stabilized therein as humus. Moreover, rising temperatures tend to hasten plant growth and to prolong the growing season in regions where growth is normally inhibited by cold weather. Such processes tend to moderate the greenhouse effect. On the other hand, rising temperatures may exceed optimal levels for some plants in some regions, thus restricting carbon assimilation, and may also hasten decomposition of organic matter, thus tending to exacerbate the greenhouse effect. Whether the positive feedbacks are likely to outweigh the negative feedbacks or vice versa will depend on site-specific conditions as well as on human intervention. In any case, the change in the temperature, which generally entails

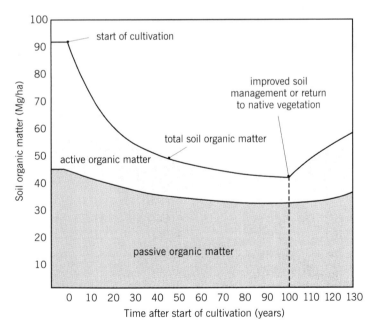

Fig. A.4. Changes in soil-carbon content after initiation of agriculture and after reforestation (schematic). (Dubbin, 2001).

change soil moisture, is certain to affect the content and turnover rate of soil organic matter.

Current models of soil organic-matter dynamics generally divide it into three categories that differ in their turnover times, from "fast" (most labile) to "intermediate" to "slow" (most stable, or recalcitrant). Other models recognize only two categories: an "active" (with short turnover time) and a relatively "inert" (with very long turnover time) fraction. As pointed out by Powlson et al. (1996), short-term studies of soil organic carbon[2] under warming may be misleading insofar as they fail to detect the response of the large "slow" reserve of soil carbon, especially as climate change is likely to affect not only the temperature *per se* but also the nutrient and moisture regimes that affect the rate of organic matter decomposition. All this serves to emphasize the importance of the stable organic carbon fraction in influencing the potential impacts of climate change on soil and vegetation. Soils that contain only a small fraction of stable organic carbon (e.g., some soils of the humid tropics) may well be the most prone to loss of productivity in a warming climate.

The process of erosion entails the removal of topsoil, the part of the soil profile that is generally most enriched with organic matter. The carbon contained in the sediment thus removed is carried by streams and either outgassed as CO_2 or CH_4 or deposited in coastal or other valley wetlands. In such wetlands (variously called bogs, fens, swamps, or marshes) organic matter may accumulate as peat even while its anaerobic decomposition releases some methane. However, whenever the land is drained of excess moisture, the organic matter stored in such wetlands tends to decompose rapidly and to emit carbon dioxide. As for the eroded upland soils, they may lose productivity permanently and become degraded, or, in some cases, be able to recover their original organic matter content and productivity by absorbing additional C from the air via photosynthesis by reestablished vegetation.

To obtain a realistic evaluation of the potential to store carbon in soils, we must consider not only the ultimate biophysical saturation of the soil reservoir, but also the socioeconomic and environmental constraints that exist in practice and that are specific to each location. Much will depend on the market costs and rewards of alternative practices, including the economic incentives instituted by policy makers to encourage C sequestration, as they are perceived and adopted by land managers.

Carbon dynamics in agricultural systems are affected by the balance between emission and uptake processes. The emission processes include:

1. Plant respiration;
2. Oxidation of organic carbon in soils and crop residues;
3. Use of fossil fuels by agricultural machinery; and
4. Use of fossil fuels in the production of inputs (fertilizers and pesticides).

The potential of a soil to sequester carbon is intimately associated with the content and nature of its clay fraction. Sandy soils, which tend to be

---

[2] Organic matter in the soil is a mixture of compounds of various compositions, usually expressed in terms of their carbon content. When oxidized to form CO_2, one atomic mass of carbon (12) combines with two atomic masses of oxygen (16+16=32) to form a molecule with a molecular mass of 44. One ton of soil carbon is thus equivalent to 3.67 tons (44/12) of carbon dioxide.

well-aerated, generally retain little organic matter. Clayey soils, on the other hand, tend to form strong physicochemical bonds between the active surfaces of clay particles and the organic macromolecules of humus, which thus become resistant to decay. Moreover, clayey soils tend to form tight and water-resistant aggregates, the interiors of which restrict aeration and further resist decay. That is one reason why disruption of soil aggregation by tillage causes not only deterioration of soil structure *per se* but also more rapid decomposition of soil organic matter.

Conventional tillage is defined as the mechanical manipulation of the topsoil that leaves no more than 15% of the ground surface covered with crop residues. In contrast, no-till management is defined as the avoidance of mechanical manipulation of the topsoil so as to leave it undisturbed and covered with surface residues throughout the period from harvesting the prior crop to planting the new crop.

Best agricultural practices can result in a net augmentation of soil carbon and in enhanced productivity due to better soil structure and soil moisture management. The relevant practices include precise and timely applications and spatial allocation of fertilizers, use of slow-release fertilizers, prevention of erosion, shortening or eliminating fallow periods, use of high-residue cover crops and green-manure crops, and minimized mechanical disturbance of soil; e.g., zero tillage. Altogether, such practices may lead to substantial restoration of the soil's organic carbon content where it had been depleted. In some special cases, it might even be possible to store more carbon than had originally been present in the "virgin" soil. Where the soils had been severely degraded and their agricultural productivity greatly impaired, they may be converted to grassland or afforested so as to serve as carbon sinks. The overall potential for carbon storage depends in each case on such factors as climate, type of vegetation, topography, depth and texture of the soil, past use (or abuse), and current management.

The crucial condition for effective soil carbon management is the ability to measure its effects on the soil. There is as yet no universally appropriate and economical method for the quantitative determination of soil-carbon dynamics over space and time in representative areas. The seemingly simple way is to extract samples from the soil and measure their organic carbon content directly, but that method is extremely laborious, destructive, and imprecise, owing to the inherent spatial variability of soil properties in general and of their carbon content in particular. Other possible approaches include the use of chambers for spot measurements of gas fluxes through the soil surface and micrometeorological monitoring of field-scale gas exchanges between the soil and the atmosphere. The problems with the use of chambers are that numerous labor-intensive measurements are needed to obtain a statistically reliable characterization of the temporally (as well as spatially) heterogeneous processes during the diurnal and annual cycles in the field, and that the measurements themselves involve artificial interference with the process on a local scale. Alternative methods, based on aerial monitoring of gas concentrations and movements above the ground either from towers or from aircraft, integrate fluxes over larger areas, but are subject to limitations due to the costs of the instrumentation and the complexity of analyzing the results.

BOX A.4   Conventional vs. Conservation Tillage

Conventional tillage buries plant residues and pulverizes the soil's aggregates, exposing their inner surfaces to aeration and thus stimulating microbial breakdown of their organic-matter. In contrast, when crop residues and cover-crops are left on the soil surface (as in conservation tillage and especially in no-till farming), they protect the soil against erosion, enhance infiltration of rainwater, reduce evaporation, and augment the organic matter of the topsoil. (Note: Conservation tillage, also called "mulch farming," is a method of soil management that maintains organic residues over the soil surface and that minimizes mechanical manipulation of the topsoil.)

In no-till farming (the ultimate form of conservation tillage), crops are sown directly into untilled soil. This practice greatly reduces soil disturbance associated with the conventional tillage. It is, however, necessarily associated with the increased use of herbicides, the manufacture and application of which involve fuel consumption and some inevitable emissions, and the operation of which may contribute to soil compaction and produce other undesirable environmental effects. Under no-till management, soils are likely to be much less susceptible to erosion, both by water (during rainstorms) and by wind (during dry spells).

WATERLOGGED PEATLANDS

Waterlogged peatlands, such as are especially abundant in Siberia and parts of Canada, generally contain large stocks of incompletely decomposed organic matter. As large areas of peat-rich permafrost are subjected to warming, they will tend at first to emit methane (the product of microbial activity under anaerobic conditions). Later, when drained of excess water, the peat will be subject to aerobic decomposition and will tend to emit carbon dioxide. Such soils may also release organic carbon in dissolved forms in the drainage efflux.

In a warming climate, the enhanced release of greenhouse gases from the thawing permafrost is a prime example of a positive feedback, by which the global warming due to anthropogenic greenhouse-gas emissions is likely to cause the secondary release of still more greenhouse gases from drained peatlands and thus to further exacerbate global warming.

Apart from the peatlands of cold regions, about 10% of global peatlands occur in the tropical lowlands, and contain an estimated 70 Pg of carbon in deposits as deep as 20 meters. Tropical peatlands are abundant in such regions as Indonesia, Malaysia, Brunei, and Thailand. These deposits appear to have been destabilized by agricultural drainage, as well as by the occurrence of more intense droughts that seem to be associated with El Niño events. Such events may result in the spontaneous burning of peat and vegetation that may cause the emission of great quantities of carbon dioxide. As more tropical swamp forests and peatlands are drained and converted to agriculture, they will likely contribute still greater emissions of carbon into the atmosphere. Such emissions may increase still further if El Niño events become more intense or frequent under a warming climate.

Global warming and CO_2 enrichment of the atmosphere may also enhance carbon absorption. However, the difficulty of obtaining an overall global estimation of future prospects inheres in our uncertainty regarding the sequence, extent, and intensity of the processes governing the balance of absorption versus emission of greenhouse gases pertaining to specific terrestrial ecosystems, and in the difficulty of integrating that balance over spatially and temporally variable landscapes. Hence, there is no single model that includes all phenomena described, and as yet no simple way to measure the relevant phenomena reliably, economically, and comprehensively.

In principle, any increase in the productivity of vegetation requires nutrients, especially nitrogen, in addition to atmospheric carbon dioxide. Therefore it is likely that N limitation will constrain the positive response of vegetation to atmospheric C enrichment. This limitation is likely to be most acute in the humid tropics, where the soils tend to be highly leached and nutrient-deficient. Hence, the potentially positive effect of atmospheric CO_2 enrichment on plant growth should not be overestimated.

BOX A.5   Trends in Agriculture Affecting GHG Emissions

At the society level, three major driving forces are to be considered: (1) the need to supply sufficient food of good quality to an increasing population; (2) the changing preferences of increasingly affluent societies for animal proteins; and (3) technological developments in food processing, storage, and distribution.

Among the concomitant changes in the agricultural system are the following: (1) clearing of native vegetation and conversion to farmland; (2) drainage of wetlands and irrigation of arid lands; (3) intensified soil cultivation and biomass burning; (4) increased use of fertilizers and pesticides; (5) greater numbers of livestock and more manure; and (6) genetic modification of livestock and crops. All these changes tend to increase the flows and cycling of carbon and nitrogen and hence also the emissions of CO_2, CH_4, and N_2O.

The task of greenhouse-gas mitigation in agriculture cannot be a single one-time action but must be ongoing site-specific combinations of measures to be applied systematically and consistently over an extended period. Such measures cannot be imposed or policed under authoritarian control but need to be promoted via a program of public education and persuasion, reinforced by positive incentives and rewards and monitored continuously. The diffuse nature of farming, with multiple practitioners and operators, is inherently difficult to control centrally.

The treatment of agricultural wastes must be such as to turn a burdensome liability into a valuable asset, a resource serving to sustain and even improve productivity. Genetic techniques should be applied, wherever possible, toward the development of crop and livestock varieties with an inherent potential for conversion of photosynthetic radiation, water, and nutrients, to better nutrition and usable energy.

IMPROVED MANAGEMENT OF SOIL ORGANIC MATTER

Soil organic matter can be regarded as a biomembrane that filters and neutralizes pollutants, while reducing sediment loads in rivers and hypoxia in lakes and river outlets. Depletion of organic matter in soils initiates a vicious cycle of degradation, affecting food security and environmental quality. Reversing that depletion via carbon sequestration can set up a benign cycle of productivity gain. Enrichment of the topsoil with organic matter makes it less prone to compaction, crust formation, and erosion. It also improves the properties of the soil with respect to infiltration, aeration, seed germination, and plant nutrition.

Conservation management of soils aims to create synergisms among several goals: augmentation of soil organic matter; enhancement of soil fertility; promotion of environmental health and sustainability; savings of energy and production costs; mitigation of the greenhouse effect; and advancement of human welfare.

Although quantitative estimates of the role of soils in climate change vary widely, many studies suggest that improved agricultural practices aimed at boosting soil organic matter content can help to mitigate GHG emissions. Among the relevant practices are the following:

1. Efficient management of tillage, crop residues, and nutrients;
2. Conservation of soil and restoration of degraded lands;
3. Efficient use of water (irrigation, drainage, reduced evaporation);
4. Selection and breeding of better crop varieties;
5. Optimal crop rotations and inclusion of cover crops;
6. Agroforestry and improved range management;
7. Superior livestock varieties, feeds, and manure utilization;
8. Reduction of fuel use for all operations;
9. Production of energy crops (biofuels) as alternatives to fossil fuels; and
10. Enrichment of soils with the additions of organic matter.

The agricultural sector can contribute to mitigation of the GHE in three principal ways: (1) reducing its own emissions by adopting such practices as no-till plantings; (2) absorbing CO_2 from the atmosphere via photosynthesis and storing a sizable fraction of that in the soil; and (3) producing renewable sources of energy, known as *biofuels*, derived from agriculturally grown biomass.

Needed is a new paradigm of GHG-conscious, GHG-efficient farming and land management in general, based on lowered energy consumption and on increased storage of carbon in soils. Especially important is the adoption of conservation tillage and zero tillage, which not only conserve energy but also enhance soil productivity. That, in turn, can relieve pressure on marginal land, stop deforestation, and maintain ecosystem biodiversity.

BOX A.6 Complementary Tasks of Biomitigation

The two major tasks of mitigating climate change are: (1) to reduce emissions; and (2) to increase absorption and storage of potential greenhouse gases. The emissions to be reduced include those from the soil (resulting from net decomposition without sufficient replenishment of soil organic matter), from above-ground biomass (burning of vegetation and of its residues), from livestock (enteric fermentation and manure), from the operation of machinery (tillage, planting, harvesting, and transportation), and from the production and application of inputs (fertilizers and pesticides). The absorption and storage of atmospheric CO_2 can be increased by means of enhanced photosynthesis by fast-growing, year-round, evergreen, dense (preferably nitrogen-fixing) vegetation capable of producing and depositing a large quantity of organic matter in the soil. A third complementary task is now coming into prominence: (3) the sustainable production of bioenergy crops—plants yielding products (including ethanol and biodiesel) that permit utilizing solar energy via photosynthesis to reduce reliance on fossil fuels.

BOX A.7 Thawing of Frozen Soils, Desiccation of Rainforests

In high-latitude areas, as ice and snow covering the ground in winter thaws, the reflectivity of the surface will decrease and more of incoming radiation will be retained as heat to further warm the soil. Permafrost soils hold a great reserve of undecomposed organic matter, perhaps as much as a fourth to a third of all soil organic carbon. Among the most labile carbon reserves are those contained as methane hydrates in permafrost and continental shelves, soils freshly cleared of tropical forests, forests that are prone to burning, as well as frozen soils that are subject to thawing. The residues of arctic plants (sedges and mosses) have been accumulating enormous amounts of organic matter over the millennia. In the high-latitude parts of Siberia, Scandinavia, and Canada, peat deposits are many meters deep. The thawing of such permafrost soils seems likely to turn them from longtime sinks to sudden sources of atmospheric greenhouse gases and thus to exacerbate global warming. While the warming bogs are still wet, they will tend to liberate methane and nitrous oxide, and as they drain and become aerated, they will tend to emit carbon dioxide. Eventually, however, these soils are likely to be invaded by grasses, shrubs, and trees that thrive in warmer and drier conditions and begin to re-sequester carbon.

Frozen soils, primarily in northern Eurasia (Siberia and northern Scandinavia) and North America (Canada and Alaska), and to a lesser degree southern South America, contain a huge amount of organic matter, estimated to exceed 400 Pg C, that had accumulated over many millennia. As these high latitudes are expected to undergo a degree of warming well above the global average (IPCC 1995), this vast reserve of organic matter is destined to undergo rapid decomposition, with concomitant emissions of C in the gaseous forms of CO_2 and CH_4, depending on local hydrological conditions. Such vulnerable carbon reserves may increase the atmospheric CO_2 content by as much as 200 ppm beyond the expected increase from fossil-fuel combustion (FAO Policy Briefs 2006).

A partially compensating process may begin sometime after the initial thawing and as the warming continues, namely the invasion of the same (formerly frozen) high-latitude areas by vegetation from lower latitudes—grasses, shrubs, and trees. That encroaching vegetation will fix carbon and begin to re-enrich the soil with organic matter. The time-rate, relative quantitative effect, and geographic extent of these conjectured sequential processes is not known in advance.

In sharp contrast to the above is the possible desiccation of tropical rainforests such as Amazonia. This potential consequence of global warming might result in the transformation of dense forests into savanna-like communities of grasses and sparse trees. Such a change could produce substantial emissions of carbon from the soil to the atmosphere and would thus constitute a positive feedback to global warming. The changes that may take place in the tropics seem likely to be more rapid than the changes likely to occur in the tundra regions, especially if the possible desiccation is accompanied by abrupt outbreaks of carbon-emitting fires. Neither of these sets of changes can be predicted with certainty, but if they do follow a pattern that in any way resembles the scenario described above, they could result in a very different world than we know at present.

There are, however, some necessary caveats: Some of the practices aimed at increasing the efficiency of agricultural production inevitably entail increased use of energy. Among those practices are irrigation, fertilization, pest and weed control, and transportation. Moreover, intensified livestock management may well result in greater methane emissions due to enteric fermentation and the disposal of wastes. Altogether, the mass balance between emissions and absorptions of GHGs tends to be uncertain, is geographically variable, and involves tradeoffs.

Some benefits of conservation farming may diminish in time. The potential for soil organic carbon sequestration is generally finite. SOC equilibrium may be attained in several decades. Higher soil temperatures due to global warming may accelerate organic matter decomposition and hence inhibit C sequestration. Only in certain special situations may the carbon content of a given terrestrial domain be made to exceed its prior natural limit; e.g., if a naturally arid and nutrient-poor area is irrigated and fertilized artificially. The extra storage of carbon is then likely to be highly labile and vulnerable, prone to turning from an absorbent sink to an emission source if the carbon-augmenting management regime is not maintained, or if it is interrupted by the occurrence of some perturbation such as drought, flooding, or fire.

Some benefits of conservation management, however, can persist indefinitely. Reduction of fuel use due to efficient operations, especially in consequence of the adoption of zero tillage, continues as long as that form of soil management is maintained. So does the improvement of soil quality, including the enhancement of soil fertility and the control of soil erosion. Of course, the possible sustainable production of energy crops to replace fossil fuels can also be a continuing benefit.

Such benefits should not be expected to occur spontaneously. Rather, they should be promoted by means of effective policies to encourage and reward C-efficient practices. Farmers are more likely to adopt C-efficient practices if given both guidance and tangible economic incentives. Any scheme to reward carbon sequestration must be based on an effective system of monitoring the results on a continuing basis, since the gains painstakingly achieved by such practices as conservation tillage, cover crops, and residue retention can be reversed subsequently (and very rapidly) by reversion to traditional tillage, residue removal or burning, and fallowing. Research is necessary to develop locally or regionally appropriate methods of monitoring, whether by sampling or—preferably—by remote sensing.

In attempting to minimize the emission of N_2O, careful attention must be devoted to the mode of fertilizer application. The traditional, spatially uniform, over-generous application of nitrogenous fertilizers or manures at the onset of the growing season may entail large losses of soluble nitrogen by leaching and of volatile nitrogen by diffusion.

Modern precision agriculture, recognizing the heterogeneity of soils in the field, applies fertilizers preferentially where most needed, and at precisely calibrated rates (including the use of slow-release fertilizers) so as to maximize nutrient-use efficiency and to minimize nutrient losses, entailing environmental pollution, such as eutrophication of freshwater bodies. Similarly, the handling of organic manures should be such as to avoid wasteful emissions of nitrogen, as well as of carbon. Increased reliance on green-manure plants (legumes and their associated nitrogen-fixing bacteria) can also help. Finally, soil moisture management will have much to do with creating or avoiding the conditions that cause GHG emissions.

The terrestrial domain and its soils should be seen as an arena wherein dynamic processes involving energy, water, oxygen, carbon, nitrogen, and numerous other vital components are in constant exchange and interaction. Within this context, carbon sequestration in soils is a technically and economically feasible and necessary option for attenuating global warming, and it can be environmentally beneficial in many other ways as well.

APPENDIX B: THE ROLE OF SOIL IN THE GLOBAL FOOD SUPPLY

BACKGROUND

A major task of our time is to ensure adequate food supplies for the world's current population (now nearing seven billion) in a sustainable way while protecting the vital functions and biological diversity of the global environment. That task is certain to become more difficult and complex as the population continues to grow (current projections anticipate the world's population to reach 8 billion by the year 2025 and to exceed 9 billion by 2050, after which it may stabilize); and—moreover—as the impacts of human usurpation of the natural environment become increasingly severe.

The greater part of the projected population growth is expected to take place in the less developed countries of Africa, South and Central America, and parts of Asia. Production of food must increase by a commensurate amount just to maintain the present nutritional levels, and by more than that if the diet in currently deficient regions is to be improved. (According to FAO estimates, as many as 800 million people were suffering from undernourishment or malnutrition as of the 1990s). The necessary improvement is not merely quantitative (i.e., measured in per capita consumption of calories, generally derived from starchy grain, tuber, or root crops). It should be qualitative as well (i.e., based on higher nutritional standards, likely to include greater consumption of animal-derived protein).

To provide future food security, the international community must recognize the processes involved and learn to manage them while avoiding, and even rectifying, the abuses of the past.

Ever since the advent of agriculture some ten millennia ago, humanity's main method of increasing food supplies has been to expand the area under cultivation and grazing, while developing methods to improve the productivity and quality of crops and livestock. The geographic expansion of agriculture from its centers of origin to ever wider areas led to the colonization of new lands in the habitable parts of every continent. In too many cases, those activities resulted in deterioration of soil productivity, caused by depletion of nutrients, loss of organic matter, erosion, pollution, urbanization, and – in the case of irrigated soils – water-logging and salination. A vicious cycle was set in motion: As agriculture expanded to meet the needs of a growing population, the population grew still larger and so did the need for ever more agricultural land. The expansion of agriculture still continues, albeit at a slowing pace, in parts of South America, Indonesia, and Africa. However, it cannot continue much longer, owing to the scarcity of suitable new areas and the damage caused to the fragile soils and the biological diversity of the remaining natural ecosystems.

The imperative to alleviate hunger by providing food security and adequate nutrition to all the world's people must depend henceforth on our ability to intensify agricultural production and to improve its efficiency and sustainability in the most suitable soils, while alleviating pressure upon the most fragile and degraded soils so as to allow (indeed, to promote) their ecological rehabilitation. At the same time, we must avoid further encroachment upon vital natural ecosystems. Given the great diversity of soils, climates, biotic communities, human lifestyles, and socioeconomic conditions around the world, these are certain to be complex and challenging tasks indeed.

HISTORICAL PERSPECTIVE

Prior to the Neolithic Transformation and the advent of agriculture (regarded as the beginning of civilization) some ten millennia ago, the entire world's population probably totaled less than ten million. At the beginning of the common era, about two millennia ago, it may have reached 200-300 million. The population rose to nearly 1 billion by the beginning of the 19th century, just two centuries ago. At that time, an area estimated at some 400 million hectares was being cultivated, mostly by hand or by animal-drawn tools, and crops were harvested by means of sickles or scythes. Soil fertility was maintained, albeit incompletely, by the slow weathering of soil minerals and the natural rate of biological nitrogen fixation, augmented in part by the practices of manuring, periodic fallowing, and increments of river silt (in irrigated areas). The typical yields of wheat were probably of the order of 1 ton per hectare, and those of rice perhaps 1.5 tons per hectares.

By the end of the twentieth century, the world's population had grown to 6 billion, the area under cultivation had doubled, and though the productivity of the soils in wide areas had declined, overall crop production kept pace with the growth of population and even raised nutritional standards in many (but not all) regions. The progress of food production was greatly helped by the advent of chemical fertilizers, more efficient irrigation and water conservation, the use of pesticides, and genetic improvement of crop varieties. However, future progress, though necessary and possible, cannot be taken for granted. It will require increased efforts at research, extension, and application – guided by appropriate policies at the national and international levels. An essential aspect of the required effort must be directed toward the maintenance, indeed the enhancement, of soil productivity.

A very significant portion of food production is achieved with the aid of irrigation, based on the withdrawal of fresh water from rivers, lakes, and aquifers. Such withdrawals generally entail serious environmental effects and often involve inter-sectoral, interregional, or international competition or even conflict. Of the total annual withdrawals of some 3240 billion cubic meters, nearly 69% are allocated to agriculture, some 23% to industry, and only about 8% to household use. Many of the major irrigation projects are based on the construction of large dams, which often are designed to supply electricity as well as irrigation water. Such dams tend to deprive the downstream environment, cause large evaporative and seepage losses, and are themselves subject to gradual clogging by the accumulation of silt. The recent trend is to construct fewer large dams, as most of the best sites have already been appropriated and as environmental concerns have gained greater recognition.

MAINTAINING AND ENHANCING SOIL FERTILITY

Soil properties affect every crop throughout its growth cycle, from seed germination to seedling emergence and establishment, and then through the stages of vegetative growth and reproductive development to maturation and harvest. All these stages depend vitally on the temperature, moisture, aeration, nutrient status, pH, and mechanical properties of the seedbed and the rooting medium.

Especially critical is the moisture supply, first to the seedlings and then to the proliferating roots. The supply must be adequate to answer the time-variable water requirements of the growing plants (subject as they are to the evaporative demand of the atmosphere which, in turn, is dictated by solar radiation, atmospheric humidity, and wind). In a typical growing season of a crop, which may last three to twelve months, the water requirements for optimal growth and yield may vary from 500 to 1500 or more millimeters (i.e., from 5,000 to 15,000 or more cubic meters per hectare), depending on weather and vegetative density.

No less important is the nutrient supply from the soil to the roots. Plant requirements are highest during the period of major vegetative growth. For maximal efficiency, therefore, fertilizer supplies should be synchronized to meet the optimal needs of the crop, taking account of the fact that the nutrient content of soils varies greatly not only between different regions but also within a single field.

The most efficient timing of fertilization can be achieved in areas that are under drip irrigation, an assembly in which the nutrients can be injected at a variable rate into the water supply and provided directly to the rooting zone. In the much larger areas where no such irrigation is available, the fertilizers should best be applied in slow-release form, or in gradual doses (i.e., in small amounts at the start of the growing season and in larger amounts during the period when plant requirements are highest). In addition to varying the fertilization rate temporally, it should also be varied spatially, to supply nutrients where they are needed most. By this means, fertilizers are not wasted by excessive application to where the soil is not deficient (the excess application often causing the pollution or eutrophication of streams, ponds, and aquifers), but are targeted specifically to when and where they are needed most. Equipment is now available (with geographic positioning systems) to allow the fertilization process to adjust to the site-specific needs of plants growing in spatially heterogeneous fields.

The nutrients required by plants and supplied as fertilizers are generally classified into two categories: (1) macronutrients, including nitrogen, phosphorus, potassium, calcium, magnesium, and sulfur, and (2) micronutrients, including iron, manganese, copper, zinc, boron, molybdenum, chlorine, and nickel. Other beneficial elements may include cobalt, sodium, and silicon.

The amount of nutrients available in a unit volume of soil varies greatly from one soil type or region to another, depending on the particular minerals present in the soil and their degree of weathering, the content of organic matter and its degree of decomposition, and the balance of nutrient additions (from the atmosphere or from aeolian or alluvial depositions) versus extractions due to leaching, erosion, volatilization, or removal by agricultural harvesting. The major nutrients in the soil, which occur as ions in the soil solution, include NO_3^-, NH_4^+, $H_2PO_4^-$, HPO_4^{2-}, K^+, Ca^{2+}, Mg^{2+}, and SO_4^{2-}.

In most agricultural soils, soil moisture is a rather dilute solution. As nutrients are absorbed from the soil solution in the immediate proximity of active plant roots, they create osmotic gradients in the soil that induce the diffusion of additional nutrients from the farther reaches of the soil toward the roots. At the same time, the extraction of nutrients induces the release into solution of more nutrients from soil minerals as well as from the continually decomposing organic matter present. All the while, the roots of a growing plant extend and proliferate so as to tap into ever newer regions of the soil profile.

In the course of this process, plant roots may be greatly aided by the presence and activity of rhizospheric microorganisms that stimulate, and in turn are stimulated by, the exudation of mucigels (gel-like organic compounds) from the roots that serve to mediate between the roots and the soil and evidently facilitate the uptake of both water and nutrients. Such symbiotic microorganisms, generally fungi, are known as mycorrhizas. The symbiosis is such that the mycorrhizas obtain their sugar requirements from the roots of plants while facilitating the uptake of both water and nutrients from the soil by the roots.

BOX B.1   Vital processes and functions performed by the soil:

1. Provides anchorage and sustenance (including oxygen and soluble nutrients) for photosynthetically active green plants.

2. Hosts a diverse community of detrivores and decomposers (meso- and micro-flora and fauna), as well as nitrogen-fixing organisms.

3. Accumulates (stores) organic residues of plants and animals in more-or-less stable forms, including humus (which, in turn, helps to sustain soil fertility).
4. Detains, transmits, filters, decomposes and/or recomposes mobile-immobile chemical constituents (solutes and precipitates).
5. Degrades toxins (including pesticides residues) and pathogens, and transmutes their decomposition products into nutrients.
6. Accumulates or releases various gases (including oxygen and the greenhouse gases CO_2, CH_4, and N_2O) in exchange with the atmosphere.
7. Mediates between meteoric water (precipitation) and aquifers via the processes of infiltration, retention, deep percolation, and groundwater recharge.
8. Absorbs, releases, and transmits energy in various forms (radiative, thermal, latent) in exchange with the atmosphere, substrata, and biota.
9. Moderates diurnal and seasonal temperature changes and thus buffers against adverse temperatures).
10. Buffers chemical reactions affecting potentially adverse changes in acidity, alkalinity, salinity, and the diffusion or osmosis of solutes.

IMPORTANCE OF SOIL ORGANIC MATTER

The presence of organic matter in the soil, per se, is not an essential requirement for plant growth. Given an optimal supply of water, oxygen, and soluble nutrients, plant roots can obtain their requirements and grow well in a soil devoid entirely of organic matter (as in a pure sand, if so supplied), or even in a solution without any soil at all (a method of growing plants called *hydroponics*). Under natural conditions, however, such an optimal environment for roots seldom exists. For soils to grow well in a real soil environment, the soil must be favorable chemically (with an optimal concentration and balance of nutrients and an absence of toxic factors), physically (with good aeration and moisture, and without mechanical impedance), and biologically (with beneficial organisms and without pathogenic agents).

The presence and state of organic matter in the soil generally helps to provide that favorable combination of conditions. Organic matter improves and stabilizes soil structure (which in turn facilitates infiltration, aeration, and root development), it serves as a growth medium and energy source for beneficial microorganisms and as a reservoir of plant nutrients that are made available gradually, and it buffers the root zone against deleterious conditions of acidity or toxicity. Especially important is the organic matter content of the topsoil that influences such vital processes as infiltration, aeration, soil temperature, and seed germination and establishment.

The amount and quality of organic matter in the soil are not constant, but subject to continual change, depending on the balance of supply (made available naturally by fresh residues of plants and animals, or artificially by manuring) versus decomposition or removal by erosion. The same principle applies to nutrients in the soil, added by the mineralization of organic matter (and, in the case of nitrogen, by the fixation of elemental nitrogen from the atmosphere) and removed by the harvesting of agricultural yields as well as by leaching and volatilization. When losses of nutrients exceed the inputs, soil fertility inevitably diminishes, and after a time may be severely depleted (as, for example, in the case of soils in the humid tropics, where organic matter decomposition is rapid and soluble nutrients are strongly leached). To maintain soil structure and soil fertility, therefore, an annual re-supply of organic matter and nutrients is generally necessary.

AVOIDING OR AMELIORATING SOIL DEGRADATION

Soil degradation is defined as the decline of the soil's productivity (or productive potential) due to one or several of the following processes: erosion of topsoil by water or wind, net loss of nutrients (that may be compensated by the application of fertilizers), development of unfavorable biochemical conditions (e.g., acidity, alkalinity, salinity, pollution, toxicity), or deterioration of the soil's physical condition (such as may be due to pulverization, compaction, waterlogging, or colloidal dispersion). Soils that can resist or overcome degradation spontaneously, or that can be readily treated so as to restore productivity, are said to be *resilient soils*. Such soils should be protected, and – where damaged – ameliorated. On the other hand, soils that tend to deteriorate practically irreparably (as, for example, highly erodible shallow sandy soils in arid regions, or acid-sulfate soils in tropical regions, or coastal soils subject to saltwater inundation) should probably be taken out of production and allowed to revert to their natural conditions.

Controlling Water Erosion

Water erosion typically begins as raindrops strike the bare soil, slake down the aggregates, and detach and splash the particles, which then tend to clog the wide pores at the soil surface. Consequently, as rainfall intensity exceeds the diminishing infiltration rate, surface water excess accumulates, forming puddles, and – if the land surface is not perfectly level – initiating overland flow (runoff). As it gathers speed running downslope, the overland flow becomes increasingly turbulent and erosive. It scours the loosened surface, cutting into it and forming rills that may eventually become deep gullies.

The erosivity of a rainstorm depends on its intensity (volume of rain falling on a unit area per unit time), its duration and total amount, and the size distribution of its raindrops (which determines their kinetic energy). The erodibility of the soil depends on its texture and structural stability, as well as on its infiltrability. Finally, the amount of water erosion (a function of both rainfall erosivity and soil erodibility) also depends on the steepness and length of the slope. In densely vegetated soils, the direct impact of raindrops is normally intercepted by the plants, which thereby protect the soil against water erosion. In most agricultural systems, the soil surface is periodically bared of vegetative cover, so it becomes particularly vulnerable to erosion by rain. Under proper management, the surface should be covered by a protective mulch consisting of plant residues, and the slope should be minimzed by terracing and broken into sections so as to prevent overland flow from accelerating as it runs downslope.

Controlling Wind Erosion

The degradation of soils in semiarid areas has been called desertification[1]. A major mechanism of such degradation is wind erosion. It typically occurs in dry regions after the soil is bared of vegetation or its residues and is pulverized or loosened by trampling or by tillage. Gusts of wind then pick up particles and either carry them aloft (when the wind is strong and the particles are small), or merely bounce them over the surface, where the striking particles knock against other particles that in turn bounce downwind (an action called *saltation*).

The erosivity and carrying capacity of wind increase exponentially with air speed, so a strong wind can literally deflate the entire loosened topsoil in a few hours or days. Wind erosion (much like water erosion) is greatly retarded by the presence of a protective layer of vegetation or its residue (mulch) at the surface, and greatly accelerated by the absence of such protection. Wind erosion can also be diminished by means of shelter belts, which are barrier (constructed fences or planted rows of trees) set perpendicular to the prevailing direction of the wind and at suitable intervals of space, so as to lower the speed (and hence the erosivity) of the wind blowing over the land surface. Wind erosion can also be reduced if the soil surface is kept in a cloddy condition rather then pulverized excessively.

Controlling Soil Acidity and Soil Pollution

Soil acidity is likely to occur where the soils are highly leached of such cations as calcium and magnesium, either due to the absence in the soil of minerals containing such cations or due to the infusion of acid rain or of industrial acids (via pollution from smoke or sewage disposal). Soil acidification may also be induced by the application of nitrogenous fertilizers containing urea, ammonia, or ammonium salts that, on oxidation in the soil by microorganisms, may produce nitric acid. One effect of soil acidity is to enhance the solubility of aluminum (from aluminosilicate soil minerals), perhaps to toxic levels that exceed the tolerance of some plants. The standard way to counter soil acidity is to add lime (e.g., pulverized limestone, containing calcium carbonate).

An important property of soils is the ability to adsorb, decompose, or otherwise neutralize chemical pollutants that might pose a danger to biota and to environmental processes in general. Included among potential sources of pollution are sewage sludge and various industrial waste products. Particularly hazardous are such metals as cadmium, chromium, copper, nickel, lead, and zinc. Such pollutants can be placed in so-called landfills, which are bodies of soil isolated by barriers from the open environment. In such sites, the hazardous materials are encapsulated in various forms and rendered harmless. Various kinds of landfills or depositories have been designed and established in many places. Intended to isolate the pollutants. Some of these sites do, however, permit the leakage or volatilization of harmful agents and must therefore by carefully monitored (especially if they contain persistent toxic or radioactive substances).

---

[1] Desertification is a process whereby an area that originally hosted a vital community of plants and animals is so damaged and its life so decimated as to resemble a desert. The mistreatment or mismanagement that can cause desertification generally includes the initial destruction of the native vegetation; followed by excessive tillage or overgrazing, causing soil erosion.

BOX B.2   On Eutrophication:

Ill-timed, poorly distributed, and excessive use of fertilizers not only involves economic losses but also poses the risk of environmental damage due to eutrophication. That term applies to the drainage of fertilizers (especially nitrogen, which is leached most readily, but also phosphate that is attached to eroded sediments) into bodies of fresh water (streams and lakes), enriching the waters with nutrients and causing the unnatural proliferation of certain forms of life that disrupt the preexisting balance of the indigenous biotic community. Most typically, eutrophication causes algal blooms that make the water of ponds turbid, and – as the algae die and decompose – deprive the water of oxygen (a condition called anoxia). The typical result is a massive fish-kill. A body of clear water, teeming with fish, can thereby turn into a putrid, slimy swamp.

AGRICULTURAL WATER MANAGEMENT

Much can be done to improve the efficiency of rainfed agriculture. The first task is to maximize the infiltration of rainfall, thereby avoiding runoff and the attendant loss of both water and topsoil. Equally important is the task of minimizing evaporation from the soil surface, by maintaining a stubble mulch over the soil surface, and of reducing transpiration by weeds, via the judicious use – where necessary – of non-inverting tillage or of herbicides.

In many semiarid regions, fallowing can be practiced to store water from one season to the next. Other means to improve the efficiency of rain-water utilization are the timely planting of drought-tolerant varieties at the appropriate density, the provision of supplementary irrigation where possible, and the optimization of all other conditions that affect growth (such as nutrient supply and pest and disease control).

Similarly, there are proven ways to improve the water-use efficiency of irrigated crops. Irrigation accounts for two-thirds of the global water requirements. It produces over a third of global agricultural production on less than one sixth of the total cultivated land. Hence one hectare of well-managed irrigated land can produce as much as three or four hectares of non-irrigated (rainfed) land. In view of the fact that most of the irrigated land is still managed very inefficiently, there is obviously great latitude for improvement.

Irrigation projects generally involve various water losses, including evaporation from storage dams and ponds, seepage from the bottoms and sides of unlined (earthen) canals and ditches, transpiration by weeds, and deep percolation beyond the reach of crop roots that also entails loss of soluble nutrients. Altogether such losses may amount to half or more of the volume of water allocated to an irrigation project. Efficient irrigation strives to minimize each of these losses; e.g., by lining channels with concrete or by conveying water through closed conduits (pipes), by eradicating weeds, and by applying irrigation in the precise amount and timing to best effect. The task of improving the efficiency of water use in irrigation is becoming increasingly vital as populations keep growing and as water resources are limited and subject to competing demands.

The conservation and the efficient and sustainable use of soil and water resources are sine qua non requisites of agricultural progress. This principle applies both to the more intensive utilization of existing agricultural lands and to the development of new lands.

BOX B.3   The Challenge of Extending the Green Revolution

The great gains made as a consequence of the so-called Green Revolution since the late 1960s were achieved thanks to a combination of advances, including the development of new varieties of grain crops with superior yielding potential, along with the increased use of fertilization and irrigation. Those gains, however, cannot be taken for granted in the future. Not only is the population continuing to grow, but – perhaps more ominously – the water supplies upon which crop yields depend are dwindling in many regions, as are the energy supplies required for modern agricultural production. The threats are due to several simultaneous processes: increasing demands for water by industries and municipalities, silting of dams, falling water tables due to overdrawn aquifers, and – finally – global climate change (including the melting of glaciers that feed some of the world's most copious rivers and the increasing aridity of entire regions). All of these factors emphasize the vital importance of improving the efficient and sustainable use of soil, water, and energy resources.

GLOSSARY

Acid rain: Rainfall that has been acidified by the dissolution of gases in the atmosphere. They natural presence of carbon dioxide in the atmosphere normally causes the pH of rainfall to be about 5.6 (slightly acidic). The anthropogenic emissions of oxides of nitrogen and sulfur further acidify the rain falling over large areas downwind of major industrial and urban centers. Acidified rain may cause damage to many forms of life, as well as to buildings and monuments.

Actinomycetes: A class of soil microorganisms intermediate between bacteria and fungi. They commonly form intricately branched thread-like mycelia that enmesh roots and extend into fine pores in the soil.

Adhesion: The attraction between substances or particles, acting at surfaces of contact between them, tending to cause them to adhere to one another.

Adsorption: Attachment of a substance to the surface of a body, impelled by forces of adhesion (e.g., adsorption of water molecules or of cations by clay particles).

Advection: Exchange of energy and matter between a field and its surroundings (e.g., the transport of heat and vapor into or out of a field by winds).

Aeolian parent material: Sediments deflated and transported by wind, generally from arid or semiarid regions. Where the material is deposited, soil formation begins. Such is the material called *loess*, which is particularly prevalent in areas of northwest China, the Ukraine, and the Great Plains of North America.

Aeration: Exchange of gases between the soil's air phase and the external atmosphere, supplying oxygen to roots and microorganisms in the soil and removing carbon dioxide (the product of their aerobic respiration).

Aggregate: Assemblage of particles bonded together. In the soil, aggregates are stabilized by organic matter (humus) or by various inorganic cementing agents.

Agronomy: The science and practice of crop production in the field, including methods of soil management.

Air encapsulation: Entrapment of air in bubbles or in bypassed pores within a nearly saturated soil.

Albedo: Reflectivity of a surface to short wave radiation (i.e., the fraction of the incoming solar radiation flux that is reflected rather than absorbed). The albedo of bare soils ranges between 0.1 and 0.4, depending on color, surface roughness, and wetness.

Algal bloom: Proliferation of algae in ponds and lakes due to enrichment with nutrients (eutrophication), generally from agricultural or urban runoff. Algal blooms may cause fresh-water bodies to become turbid and anoxic (deprived of oxygen).

Alkalinity: High concentration of sodium ions in the soil's exchange complex. It generally causes an increase of pH, dispersion of clay and humus, and breakdown of aggregates, thus reducing soil permeability.

Allophane: A class of clay minerals that are amorphous rather than crystalline. Such minerals are especially prevalent in soils formed of volcanic ash.

Alluvium: A deposit of fluvial sediments, typically conveyed and deposited by streams. Large accumulations of alluvium occur in river valleys and in some coastal plains.

Aluminosilicate clay minerals: Clay minerals in which the crystal lattice consists of alternating layers of alumina and silica ions in association with oxygen or hydroxyl ions.

Amendment, soil: A material, such as lime or gypsum, that is applied to the soil to improve its condition (e.g., to neutralize acidity or alkalinity) and its productivity.

Ammonification: The transformation of nitrogen-containing organic compounds into ammonia by biochemical processes.

Anaerobic soil conditions: Restricted soil aeration, resulting in chemical and biochemical reduction reactions (including denitrification; generation of methane, ethylene, and hydrogen sulfide; and reduction of manganese from the manganic to the manganous state as well as of iron from the ferric to the ferrous state).

Aquifer: A porous geological formation that contains and transmits groundwater to natural springs or to wells that supply water for human purposes.

Arid regions: Regions in which the natural supply of rainfall is below the potential evaporation rate during much of the year, hence crop plants are likely to suffer water deficits and physiological stress.

Autotrophs: Organisms (including green plants and cyanobacteria) capable of synthesizing their own food from inorganic sources, utilizing solar or chemical energy.

Bioremediation: Use of biological organisms (such as toxic-absorbing microbes or aquatic plants) to absorb and neutralize pollutants and thus decontaminate soils.

Buffering capacity: The capacity of a soil to resist changes in acidity or alkalinity. It is due largely to the presence and activity of humus and clay.

Bulk density: Mass of soil particles per unit bulk volume (which includes the combined volumes of particles and pores), indicative of the compactness or porosity of a soil body.

Buried soils: Soils formed in an earlier period, subsequently covered by alluvial or Aeolian depositions. In some cases, a series of soils may form one atop the other, as each soil in turn is covered with sediment that is then subject to a new cycle of soil formation.

Caliche: A layer of soil cemented by the accumulation of calcium and magnesium carbonate, generally occurring in the B-horizon in lime-rich soils in arid regions.

Capillarity: The condition of water being drawn into the narrow ("capillary") pores of the soil, where it is generally at a sub-atmospheric pressure called tension or suction.

Capillary fringe: The zone just above the water table that is still saturated, though under sub-atmospheric pressure.

Carbon cycle: The cyclic series of processes by which atmospheric carbon is absorbed and fixed into living organisms by photosynthesis or chemosynthesis, then returned to the atmosphere by respiration and decomposition.

Carbon sequestration: The accumulation of organic carbon in the soil, resulting from the addition of plant and animal residues and their partial decomposition to form stable humus.

Catena: A succession of soils derived from similar parent material under similar macro-climatic conditions, but differing positions in the landscape (e.g., plateau, upper slope, lower slope, valley bottom, etc.), hence reflecting the influence of microclimatic variables (moisture and temperature).

Cation exchange capacity: the total capacity of a soil, per unit mass, to adsorb and exchange cations. It is generally expressed in terms of milliequivalents per 100 grams.

Cementation: Precipitation within the soil profile of calcareous, siliceous, or ferruginous compounds that bind particles into a hardened and relatively impervious mass.

Characteristic curve of soil moisture: A graph relating soil wetness to matric suction. If measured in desorption, it is the moisture retention (or moisture release) curve.

Chlorosis: A condition of plants, generally caused by nutrient (e.g., nitrogen) deficiency, generally resulting in reduced vitality and pale green (or yellowish) coloration of leaves.

Clay: The colloidal fraction of mineral soils, consisting of particles smaller than 2 microns in diameter.

Clay minerals: A group of "secondary" minerals that form in the soil as a result of the partial dissolution of primary minerals and their re-precipitation during the long process of pedogenesis. The most prevalent clay minerals are the aluminosilicates.

Claypan: A clay-enriched layer occurring at some depth in a soil profile, that is relatively tight and impermeable to water, and sometimes impenetrable to roots.

Cohesion: The internal bonding of molecules or particles of a particular substance.

Colloids: Materials that are composed of very fine particles having a large surface area per unit mass, such as aerosols in the atmosphere or clay in the soil.

Colluvium: Sediment accumulating at the base of mountains, resulting from the gravitational slumping soil material from destabilized slopes.

Compaction of soil: Densification of an unsaturated soil body by the reduction of its air-filled porosity, generally resulting from the application of compressive forces (pressure).

Compost: Partially decomposed organic material (e.g., plant and animal residues) that can be applied to soils to improve their fertility and physical condition.

Consolidation of soil: Densification of a water-saturated soil, as in the case of a heavy structure resting on compressible clay (a frequent cause of subsidence).

Contact angle (of a liquid on the surface of a solid): A measure of the affinity between a liquid and a solid. A drop of water tends to spread over a hydrophilic (wettable) solid, so the contact angle is acute (and may even be zero). On the other hand, water tends to "ball up" when placed over a hydrophobic water-repellent) surface, so its contact angle is obtuse (and may even approach 180 degrees).

Convection: The mass flow of a fluid in response to a driving force acting on the entire body of the fluid (such as a pressure or gravity gradient).

Cover crop: A crop that produces a dense stand, grown between rows of trees in orchards, or between regular cropping seasons in fields, aimed to provide vegetative cover for soil protection and enrichment.

Crop-response function in irrigation: Functional relationship between the yield of a crop and the amount of water applied to it. When plotted as a curve, this function is in general sigmoid in shape, its slope at each point representing the incremental yield response to a unit volume of added water.

Crop rotation: A series of different crops grown in succession in a particular tract of land, to better maintain soil fertility. Most crop rotation programs include a leguminous crop at every cycle (once every two, three, or more years) so as to enrich the soil with nitrogen.

Crust formation: Development of a dense, hard, and relatively impermeable layer at the soil surface, due to the breakdown of soil structure there. A surface crust may inhibit aeration, infiltration, and the emergence of germinating seedlings.

Cryoturbation: The action of repeated cycles of freezing and thawing in cold-region soils, causing the soil to swell, shrink, and churn repeatedly.

Darcy's law: The rate of flow of a water in soil (volume passing through a unit area per unit time) is proportional to the hydraulic gradient. The proportionality coefficient is called the hydraulic conductivity.

Decomposition: The chemical or biochemical breakdown of complex compounds into simpler ones, such as the decomposition of organic compounds (e.g., plant residues) in the soil by microorganisms that release soluble nutrients for the renewal of plant growth.

Deflation: Removal of loose soil particles from a bare soil surface by blowing wind; a process of wind erosion induced by the denudation and pulverization of the soil surface mainly in arid regions.

Degradation: Deterioration of soil productivity by various processes such as erosion, organic-matter depletion, leaching of nutrients, compaction, breakdown of aggregates, waterlogging, and/or salination.

Denitrification: The reduction of nitrates in the soil to nitrite, nitrous oxide, or nitrogen gas in anaerobic soils.

Density of soil solids: Mass of soil solids per unit volume of the solids (excluding the volume of pores). The density of most mineral soils is about 2.65 grams per cubic centimeter. The presence of organic matter in the soil lowers its solid density.

Desertification: Degradation of land in semiarid regions, caused by destruction of the vegetative cover by overgrazing or tillage, depletion of organic matter and nutrients, breakdown of soil structure, and erosion. An originally productive area may thus come to resemble a desert.

Desert pavement: The formation of a layer of gravel and stones on the soil surface of desert areas, due to the deflation of finer particles by wind and water erosion.

Diffuse double layer: A phenomenon characteristic of hydrated clay particles, whereby the negative electrostatic charges on the clay surfaces are countered by the positive charges of a spatially diffuse swarm of cations in the surrounding aqueous solution.

Diffusion: The net movement of molecules of different substances relative to one another (due to their random thermal motion) in a mixed gaseous or liquid medium, tending toward equalizing the concentrations of all components throughout the medium.

Drainage: Outflow of water from the soil, either naturally or artificially. Surface drainage refers to the downslope flow of excess water from the soil surface. Subsurface drainage (groundwater drainage) refers to the removal of water from within or below the soil, generally lowering the water table.

Dryland farming: Production of crops in semiarid regions by reliance on rainfall alone, without irrigation. Its proper practice requires care to conserve soil and water by promoting infiltration, minimizing evaporation, and preventing erosion.

Earthworms: Animals that burrow into the soil and ingest organic residues. In the process, they churn and mix the soil, redistribute organic matter within it, and open channels for aeration, water flow, and the rooting of higher plants. *Lumbricus terrestris* is the most common earthworm.

Ecosystem: A community of mutually dependent plants, animals, and microorganisms that share the same habitat and that exchange and recycle energy and vital materials.

Electrical conductivity: The ability of the soil solution to conduct electricity, a property that depends on the electrolyte content (salts) of the soil solution. It is used as an indicator of soil salinity.

Eluviation: Leaching of soluble or suspended material from a soil layer by percolating water.

Emissivity: The property of a surface to emit electromagnetic radiation, in proportion to the fourth power of its temperature. The proportionality factor is unity for a perfect "black body" and less than that for most other surfaces.

Energy balance of a field: An expression of the energy conservation law, whereby the sum of all energy inputs minus all energy outputs of a field (including radiant energy, sensible heat, and latent heat) equals the change in energy content of a given body of soil over a specified period of time.

Eolian sediments: Masses of fine particles (sand and silt) deposited by wind. Extensive areas of sand occur in the deserts of North Africa and Central Asia. Finer-grained deposits (called *loess*) occur in North America, Eastern Europe, China, and the Middle East.

Erodibility: The vulnerability of a soil to erosion, which depends on the texture, structure, looseness and roughness of the soil surface.

Erosion: The detachment and removal of soil particles from the soil surface, by running water or by wind.

Erosivity: The erosive power of rainfall, running water, or wind. In the case of rainfall, it depends on rainfall intensity and duration. In the case of running water, it depends on velocity and turbulence of overland flow. In the case of wind, it depends on wind speed and direction.

Essential nutrients: Chemical elements that are required for the normal growth, physiological functioning, and yield of crop plants.

Eutrophication: Enrichment of water bodies with nutrients (such as nitrogen and phosphorus). In surface waters, it may lead to proliferation of algae and to the oxygen-deprivation of fish. In aquifers, it may render the water unsuitable for drinking.

Evaporativity: The power of the atmosphere to vaprize and remove water from a moist surface, at a rate depending on ambient temperature, humidity, and wind speed.

Evapotranspiration: The transfer of water from its liquid state in the soil-plant system to the atmosphere as vapor. It includes evaporation from the soil surface and transpiration from the canopies of plants. The two processes are simultaneous and interactive, and their relative proportions depend on density of plant cover.

Exchangeable sodium percentage (ESP): The percentage of the soil's exchange capacity that is occupied by sodium ions. When ESP exceeds 15%, the soil is considered to be an alkali soil.

Failure stress: The value of stress (compressive or shearing stress) at which a body (of soil) collapses or fractures. It is a relevant property of a soil related to its ability to bear traffic or to serve as foundation for structures.

Failure: The reaction of a soil body to stresses that exceed its strength, generally leading to loss of cohesion or structural integrity by such modes as fracturing, slumping, plastic yielding, or liquefaction.

Fauna: The community of animals, both large and small, that inhabit a given habitat or ecosystem.

Fertigation: The injection of soluble fertilizers into the water supply so as to fertilzie a crop while irrigating it. Fertigation is often practiced with drip irrigation.

Fertility of soil: The presence in the soil of nutrients in forms available to plants (particularly crop plants) and in amounts adequate to promote their optimal growth and yield.

Fertilizers: Substances applied to the soil to supply nutrients needed for crop production.

Field capacity: An empirical measurement supposed to represent the soil profile's ability to retain water after the process of internal drainage has ceased. It is usually measured about two days after an infiltration event. The measured value usually depends on the texture and layering of the soil profile, as well as on the initial depth of wetting.

Flocculation and dispersion: The tendency of clay particles in an aqueous suspension to either clump together into flocs or to separate from one another and thus disperse in the fluid medium, depending on the composition and concentration of the electrolytes in the ambient solution.

Floodplain: The section of a river valley that is subject to inundation when the river is in spate and overflows its banks. Floodplains are normally covered with alluvium that had been deposited over time by the river.

Flora: The community of plants, both large and small (e.g., including algae and fungi) that inhabit a given habitat or ecosystem.

Fourier's law: The flux of heat in a homogeneous body (e.g., a soil) occurs in the direction of, and in proportion to, the temperature gradient. The coefficient of proportionality is known as the thermal conductivity of the medium.

Fragipan: A hard and brittle subsurface layer that occurs at some depth in some soils. Such a layer may inhibit water percolation and roots penetration.

Fungi: Simple forms of plant life that generally do not perform photosynthesis. Many fungi are saprophytic, i.e. derive their nourishment from dead or decaying organic matter.

Gibbsite: A mineral composed of aluminum hydroxide, prevalent in the soils of humid tropical regions.

Gilgai: A clayey soil that exhibits a typical microrelief of hummocks and pits, resulting from the swelling and shrinking that occurs during successive periods of wetting and drying.

Glacial till: Material transported and deposited by glaciers, generally consisting of variable, loose, and unsorted mixtures of boulders, gravel, sand, silt, and clay.

Goethite: A mineral composed of iron oxide that is of widespread occurrence and that imparts to many soils a characteristic yellowish-brown coloration.

Greenhouse effect: Inhibition of the release of outgoing thermal radiation from the earth to space, due to the presence of certain radiatively active gases in the atmosphere. Among those gases are water vapor, carbon dioxide (whose concentration has been increasing due to the burning of fossil fuels, the clearing of forests, and the decomposition of soil organic matter), methane, and nitrous oxide.

Greenhouse gas emissions: Emission from the soil into the atmosphere of gases (such as carbon dioxide, methane, and nitrous oxide) that contribute to global warming. Such emissions can be reversed by judicious soil management, with the soil capable of sequestering (rather than emitting) carbon and nitrogen.

Green manure: A crop, generally a nitrogen-fixing legume, that is grown for the purpose of being incorporated into the soil for the purpose of enhancing its fertility.

Groundwater: The water contained in the saturated portion of the soil or the underlying porous formations, generally at a pressure greater than atmospheric.

Halophyte: A plant that is particularly tolerant of soil salinity.

Hardpan: A hardened layer within the soil profile, cemented by silica, iron oxide, or lime.

Heat capacity: The change of heat content of a unit volume of a body per unit change of temperature.

Heat conduction: Propagation of heat within a body, driven by a temperature gradient. The quantitative expression of heat condution is known as Fourier's Law.

Heaving: Lifting of the soil, as well as of structures resting on it, due to the expansion of ice lenses in cold-region soils during freezing periods.

Heavy metals: Metalic elements (often used and discarded by the chemical industry) that may accumulate in the soil to levels that are toxic to plants and animals. Among such elements are Cd, Co, Cr, Cu, Fe, Hg, Mn, Mo, Ni, Pb, and Zn.

Heterotrophs: Organisms deriving their energy from decomposition of organic compounds (in contrast with autotrophs).

Horticulture: The agricultural production of fruit, vegetable, and ornamental crops.

Humid regions: Regions in which the amount of rainfall equals or exceeds the potential evapotranspiration throughout the growing season. In such regions, plants may suffer from excess water in the soil and impeded aeration, but seldom from shortage of water.

Humus: The fraction of soil organic matter that remains fairly stable after the initial decomposition of plant and animal residues. It is usually dark-colored, occurs mostly in the surface zone (A-horizon) of the soil, and helps to stabilize soil aggregates.

Hydraulic conductivity: the ratio between the flux of water through a porous medium and the hydraulic gradient (i.e., the flux per unit hydraulic gradient).

Hydraulic head: The sum of the pressure head (hydrostatic pressure relative to atmospheric pressure) and the gravitational head (elevation relative to a reference level). The gradient of the hydraulic head is the driving force for water flow in porous media.

Hydrodynamic dispersion: The tendency of a flowing solution in a porous medium to disperse, due to the non-uniformity of the flow velocity in the array of conducting pores.

Hydrological cycle: The reciprocal exchange of water between the atmosphere and the terrestrial domain, including the successive or simultaneous processes of precipitation, infiltration, runoff, evapotranspiration, etc.

Hydrology: The study of the dynamic state, cyclic movement, transformations, spatial and temporal distribution, and interactions of water in the various domains in which it occurs, as it affects living systems in general and human life in particular.

Hydrophobicity (water repellency): The tendency of a surface or an assemblage of particles to resist wetting by water, caused by the obtuse (rather than acute) contact angle between water and the surface. Most mineral soils have a positive affinity for water, hence are hydrophilic rather than hydrophobic. However, when coated with waxy resins, soil particles may become hydrophobic and the soil as a whole may repel rather than absorb water applied to it.

Hydroponics: The production of plants in a nutrient solution without soil.

Hyphae: Thin filaments formed by fungi in the soil, such that enmesh roots and penetrate into fine pores, thus mediating between higher plants and soil regions that roots cannot reach.

Hysteresis: The dependence of an equilibrium state on the direction of the process leading to it. A prime example is the difference between the soil moisture characteristic curve measured in sorption (i.e., as an initially dry soil is gradually wetted) and in desorption (i.e., as an initially saturated soil is gradually dried).

Illuviation: Deposition and accumulation of soluble or suspended material in the lower horizon (the B-horizon) of the soil profile.

Infiltrability: The rate of infiltration that occurs when water at atmospheric pressure is applied and maintained at the soil surface. Infiltrability is relatively high when the soil is initial dry, but diminishes in time as the soil is wetted to greater and greater depth.

Infiltration: Entry of water into the soil, generally by downward flow through all or part of the soil surface.

Internal drainage: The post-infiltration redistribution of soil moisture in the soil profile, generally at a gradually diminishing rate, until it seems to cease after a few days.

Ion exchange: The adsorption of ions (particularly cations) by the surfaces of clay particles, and their dynamic exchange with cations in the ambient aqueous solution.

Irrigation: the practice of supplying water to the root zone of crops so as to permit farming in arid regions or to offset drought in semiarid regions.

Irrigation efficiency: The fraction of applied water that is effectively consumed by plants.

Isomorphous replacement: Substitution of one ion by another of similar size in the lattice of crystals. An example is the replacement of of Si with Al, or of Al with Mg in clay minerals.

Isotopes: Atomic variants of a given element, exhibiting different numbers of neutrons in the nucleus.

Kaolinite: An aluminosilicate clay mineral of the 1:1 type that is prevalent in highly weathered soils.

Lacustrine deposits: Alluvial material that is deposited in lakes.

Laminar flow: The flow of a fluid such that adjacent laminae (layers) moving at different velocities, slide over one another smoothly, without creating eddies (in contrast with turbulent flow, where swirling eddies form in the flowing medium).

Laterites: Soils typically occurring in the humid tropics, where silica tends to dissolve and leach while iron and aluminum oxides accumulate and impart a typical red color to the soil. Chunks dug up from laterites and dried in the sun tend to harden and form bricks, hence the name derived from the Latin word *later*, meaning "brick."

Leaching: Removal of solutes from the soil by the downward percolation of water. This process is desirable where it removes excess salts from saline soils, but it is undesirable where it removes essential nutrients from non-saline soils.

Lime: Acid-neutralizing material (most typically, ground limestone, containing calcium carbonate) that is applied to soils of low pH to improve their fertility.

Limiting factor: Any factor that is essential for normal plant growth (e.g., various nutrients), the deficiency of which limits growth.

Loam: A soil of intermediate texture that contains a balanced proportion of the textural fractions sand, silt, and clay.

Loess: Aeolian material, consisting mainly of fine sand and silt, that has been transported and deposited by wind. In some places (most notable in the Yellow River Basin of northeast China), the deposits are tens or even hundreds of meters thick. They are characteristically fertile but highly erodible. Where they are cut by streams or by human action, loessial deposits tend to form vertical walls that may slump periodically.

Lysimeter: A container filled with soil and embedded in a field so as to allow measurement of soil-plant-water relations. The most elaborate lysimeters are large enough to accommodate crops (even trees) and can be weighed continuously so as to monitor daily evapotranspiration and crop-water use.

Macronutrients: Essential elements that are required in relatively large amounts, including N, P, K, S, Ca, and Mg.

Macropores: Relatively large pores in the soil, generally several millimeters in width, occurring between aggregates or a cracks or fissures in the soil matrix. Some macropores are biogenic (e.g., channels formed by decaying roots or by earthworms). If open to the surface, macropores may facilitate infiltration and aeration, but may also allow greater evaporation.

Matrix of the soil: The assemblage and arrangement of the solid phase (mineral and organic components) constituting the body of the soil, and the interstices (pores) contained within it.

Mature soil: A fully developed soil (with a characteristic profile) that is at equilibrium with its physical and biological environment.

Mechanical analysis: A procedure by which the array of particle sizes of a soil is determined quantitatively by means of sieving, sedimentation, and/or microscopy.

Meniscus: The curved surface of a liquid inside a capillary pore. The curvature of the meniscus depends on the liquid-solid contact angle as well as on the liquied-to-gas surface tensin and the atmospheric pressure.

Microirrigation: The low-volume, high-frequency application of water to the soil, via narrow-orifice emitters set in tubes that convey water at low pressure. Microirrigation includes such techniques as drip, trickle, bubbler, and microsprayer irrigation.

Mineralization: The transformation of elements from organic to mineral form, due to microbial activity. A prime example is the mineralization of nitrogen from plant residues or applied manure to nitrate.

Montmorillonite (smectite): An aluminosilicate clay mineral of the 2:1 type that tends to swell upon wetting and to shrink when drying. It has an extremely high specific surface area (~ 800 square meters per gram).

Mineral soil: A soil that is composed mainly of minerals, with relatively little organic matter.

Mottling: Formation of spots of discoloration due to chemical reduction in soils that undergo periodic waterlogging and restricted aeration.

Mulch: a layer of organic material on the soil surface, consisting of plant residues, serving to enrich the soil's organic matter content and to protect it against direct raindrop impact and erosion during rainstorms, as well as against rapid desiccation and erosion by wind during subsequent dry spells.

Mycorrhizae: A symbiotic association of filamentous fungi with the roots of higher plants, serving to enhance nutrient uptake by plants.

Nematodes: small, elongated, unsegmented worms that proliferate in vegetated soils and feed on plant roots.

Nitrification: Oxidation of ammonia in the soil to nitrate, performed by aerobic bacteria.

Nitrogen fixation: Conversion of elemental (atmospheric) nitrogen to chemical forms that can be assimilated by plants, generally performed by bacteria that may be free-living or symbiotic (living in association with the roots of legumes).

Organic soils: Soils formed in waterlogged conditions (e.g., in bogs or marshes) such that allow organic matter (e.g., peat) to accumulate with only partial decomposition. Such soils may contain 50% or more of organic matter.

Ortstein: An indurated layer in the B horizon of some soils (e.g., spodosols), cemented by illuviated iron oxides.

Osmosis: Diffusion of molecules or ions of a substance in solution through a barrier (such as a biological membrane) having pores of molecular size.

Overland flow (runoff): The flow of rainwater over the surface of the ground, caused when the intensity of rainfall exceeds the rate of infiltration into the soil.

Oxidation: Chemical combination with oxygen, or – in a more general sense – a chemical reaction in which a given element loses (or contributes) one or more electrons.

Parent material: Partially weathered rock or geologic deposit that is subject to soil-forming processes and that eventually gives rise to a developed soil.

Particle size distribution: The array of particle sizes composing the soil, typically represented in graphic form as a particle-size distribution curve.

Peat: Accumulation of partially decomposed organic matter in a marsh or swamp.

Ped: A term used by pedologists to describe a unit of soil structure, i.e. an aggregate formed by natural processes.

Pedogenesis: The gradual process of soil formation from its parent material, including the dissolution and reconstitution of minerals, the migration of solutes and particles, the accumulation of organic matter, and the development of a characteristic soil profile.

Perched groundwater: an accumulation of water that is under positive pressure, at some depth in or under the soil profile, resting on a relatively impermeable layer that lies above the regional water table.

Percolation: The downward flow of water through saturated or nearly saturated layers of the soil profile.

Permafrost: Permanently frozen ground, or a frozen layer within or below the soil.

Permanent wilting point: The depletion of soil moisture to the point at which plants wilt and fail to recover even if placed in a humid atmosphere. It is considered the lower limit of soil moisture availability to plants.

Permeability: The capacity of a porous medium to permit the flow of fluids through its continuous array of pores.

pH: An index of the acidity or alkalinity of a solution; specifically, of the soil solution. A pH value below 7 indicates acidity and above 7 indicates alkalinity. The scale is logarithmic, based on the concentration of hydrogen ions.

Plow pan: A compact layer formed in cultivated soils, just below the tilled topsoil, as a result of the pressures exerted by tillage implements and tractor wheels.

Poisuille's law: The volume rate of laminar flow through a narrow capillary tube is inversely proportional to the viscosity and directly proportional to the pressure drop per unit distance and to the fourth power of the tube radius.

Pore size distribution: The volume fractions of pores of various size ranges.

Porosity: Fractional volume of voids (pores) in the bulk volume of the soil.

Potential evapotranspiration: The rate of evapotranspiration from a dense stand of an actively growing herbaceous crop (e.g., grass) that is well endowed with water. As such, potential evapotranspiration represents the maximal weather-determined rate of evaporation from a vegetated surface. It depends on such meteorological variables as solar radiation, air temperature and humidity, and wind speed.

Pressure-plate apparatus: An instrument to measure the soil moisture characteristic in the range of 1 to 15 or more bars of tension. Pressurized air is used to extract water incrementally from initially saturated soil samples resting on a porous plate.

Pore-size distribution: The array of pore sizes in a soil of a given structure. It is usually measured by the gradual extraction of water from an initially saturated sample. (In reality the pore space in the soil, however geometrically complex, is continuous; hence the notion of distinct pores is merely a convenient assumption to allow characterization of the capillary retention of water in the soil at different tension values, as if the soil were a bundle of distinct capillary tubes.)

Porosity: The fractional volume occupied by pores in the soil. It depends on the texture and the structure of the soil, and may vary from 0.3 in densely packed soils to 0.6 in loose soils.

Primary minerals: Minerals present in the original parent material (i.e., in igneous rocks) and that remained unaltered since (in contrast with secondary mineral, reconstituted in the process of soil formation).

Profile of the soil: A vertical section of the soil in situ, from the surface through the soil's sequential horizons, down to the underlying parent material.

Protozoa: Single-celled organisms that prey on bacteria, fungi, and other soil microbes.

Puddling: Mechanical manipulation of a wet soil, especially of a clayey soil that tends to be plastic when very wet, resulting in the destruction of the soil's natural aggregates. After drying, the puddled soil tends to form massive, hardened clods.

Radiation balance: The sum of all incoming minus outgoing radiant energy fluxes for a given surface. The radiation balance for a typical soil surface is generally positive during daytime (as the incoming shortwave solar radiation flux exceeds the earth's outgoing longwave radiation flux) and negative during nighttime.

Radiation: The emission of energy in the form of electromagnetic waves from all bodies above 0 degrees Kelvin. The intensity of the emitted radiation is proportional to the fourth power of the body's surface temperature. It also depends on a property of the body itself called emissivity.

Redistribution of soil moisture: Post-infiltration movement of soil moisture in a partially wetted profile in the absence of a shallow water table, during which water moves progressively (at a diminishing rate) from the initially wetted upper zonr to the unwetted lower part of the profile, and the initially sharp wetting front becomes increasingly diffuse.

Redox potential: A quantitative measure of the reducing or oxidizing power of a medium. The redox potential of anaerobic soils is usually determined by measuring the difference in potential between an inert platinum electrode and a reference electrode (such as a saturated calumel or Ag:AgCl solution).

Reduction: In chemical reactions, the counter to oxidation; i.e., the gain of electrons by the reduced reactant.

Regolith: Fragmented, unconsolidated debris, consisting of weathered stones, gravel, and soil material.

Residual soils: Soils formed in situ on the original substrate (bedrock), in contrast with soil formed of transported material.

Rheology: A branch of the science of mechanics, describing the deformation of bodies under applied stresses, and more specifically the stress-strain-time relations of deformable bodies.

Rhizobia: Bacteria that enter into symbiosis with the roots of some types of higher plants (legumes) and that are capable of converting atmospheric nitrogen into organic forms.

Rhizosphere: The zone of the soil in the immediate periphery of plant roots, a zone occupied by microbial forms that depend on the presence and functions of roots.

Richards equation: A flow equation for unsaturated soils, combining Darcy's law with the continuity equation, along with the proviso that in an unsaturated soil the hydraulic conductivity is not a constant but a function of matric potential (or of soil wetness).

Rill: A small channel scoured by rainfall in a bare soil surface of a sloping soil. Over time, where rainfall intensity exceeds infiltrability and as runoff gathers speed and erosive power, rills may cut more and more deeply into the soil and become gullies.

Root zone: The part of the soil profile that is penetrated and permeated by plant roots, and that provides them with their water, nutrients, and oxygen requirements.

Runoff: The volume of water that runs off the soil surface and flows downslope into streams during periods of intense rainfall that exceeds the soil's infiltrability. (See also: Overland flow.)

Runoff inducement: The practice of treating the soil surface on sloping ground so as to reduce infiltrability and increase runoff as well as of collecting, directing, and utilizing the runoff thus obtained. Also called "water harvesting," this practice has been applied since ancient times by societies in arid regions, to obtain drinking water and irrigation of crops.

Saline seep: An area where saline water seeps from below to the soil surface, where it evaporates and deposits salts.

Salinity of soils: The excessive accumulation of salts in the soil, such that restricts normal plant growth and causes a decline in soil productivity. Salinity affects plants by reducing the osmotic potential of soil moisture (the soil solution) and by the toxicity of specific ions (such as boron, chloride, and sodium).

Saltation: the bouncy movement of sand particles caused by wind blowing over the surface. Gusts of wind pick up protruding particles and carry them downwind in short spurts. As they fall to the ground, they strike other particles and bounce them up in sequence.

Savanna: A grassland with scattered trees, occurring typically in a subhumid to semiarid zone, intermediate between forest and semidesert.

Secondary minerals: Minerals that are reconstituted from the weathering and dissolution of primary minerals during the process of pedogenesis.

Sedimentation: The deposition of eroded particles, which takes place where the flowing water or blowing wind that had detached and transported the particles becomes quiescent.

Seedbed: A shallow, loosened layer at the surface of the soil, where seeds can be placed and where they can readily germinate and become established.

Seepage: The loss of water from a pond or reservoir by percolation into and through the soil. An alternative use of the term is to describe the emergence of water from an exposed stratum of saturated soil.

Semipermeable membrane: A membrane that permits the passage of solvent molecules but not the molecules or ions of certain solutes. When such a membrane is placed between solutions of different concentrations, osmotic effects occur. Natural membranes fulfill a vital role in the physiology of plants and animals. Artificial semipermeable membranes serve in many industrial processes, including the desalination process of reverse osmosis.

Septic tank: A container into which domestic waste (sewage) is directed, and where initial decomposition takes place. The liquid effluent can then be drained into the soil.

Sewage sludge: Solids removed from partially treated sewage, having a high content of organic matter. In some cases, it is used to treat and fertilize soils, but it must be used with great caution lest it contain toxic and pathogenic agents.

Shelterbelt: A row or strip planted with trees aimed as reducing the speed, evaporativity, and erosivity of prevailing winds.

Silt: A class of soil particles of intermediate size, between clay and sand (i.e., between the grain-diameters of 0.002 and 0.05 millimeter).

Silting: The accumulation of water-borne sediments in streambeds, reservoirs, and river valleys.

Smectite: A group of clay minerals (including montmorillonite, beidellite, and saponite) with expansive crystal lattice, high specific surface, and high cation exchange capacity.

Sodium adsorption ratio (SAR): The ratio between the concentration of sodium ions in an aqueous solution and the square root of the combined concentrations of calcium and magnesium. This ratio serves to predict the equilibrium exchangeable sodium percentage (ESP) of a soil that is to be irrigated with the given solution.

Sodic soil: A soil with a high content of exchangeable sodium, typically with poor structure owing to dispersion of clay, and hence being an unfavorable medium for plants.

Soil: The weathered and fragmented outer layer of the earth's terrestrial surface, formed initially through the disintegration and decomposition of rocks by physical and chemical processes, and influenced subsequently by the activity and accumulated residues of numerous species of microscopic and macroscopic biota.

Soil classification: Systematic definition and categorization of soil types on the basis of their distinct physical, chemical, and biological properties.

Soil conditioners: Materials used to treat soils for the purpose of improving their physical conditions (e.g., to stabilize surface-zone aggreges so as to prevent slaking and crusting of the soil surface and to promote germination and seedling establishment).

Soil conservation: Measures to protect soils against erosion and other forms of degradation (e.g., nutrient depletion, acidification, salination, and structural breakdown).

Soil erosion: Detachment and removal of soil material by water, ice, wind, or tillage.

Soil fertility: The ability of a soil to provide essential nutrients to growing plants.

Soil organic matter: Plant and animal residues in all stages of decomposition.

Soil-plant-atmosphere continuum: A concept recognizing that the field with all its components – soil, plants, and ambient atmosphere taken together – constitutes a physically integrated dynamic system in which the various flow processes involving energy and matter occur simultaneously and interdependently like links in a chain.

Soil profile: A vertical section of the soil, revealing its sequential horizons and their major characteristics, from the top (A-horizon) to the intermediate (B-horizon) to the lowermost (C-horizon).

Soil quality: The overall functioning of the soil within the ecosystem, including its ability to sustain plant and animal life, to filter and recycle water, to maintain air quality, and – where necessary – to produce food and other products for humans without deteriorating.

Soil solution: The water contained in the soil and the various substances dissolved in it, including available nutrients as well as other salts, acids, bases, and organic solutes.

Soil quality: The capacity of a soil to support biological productivity and to perform and sustain ecological functions such as recycling of energy, water, residues, and nutrients.

Soil structure: The arrangement and organization of soil constituents into aggregates that determine the spatial configuration and porosity of the soil matrix.

Soil taxonomy, U.S. system: The classification of soils into groups, based on their respective characteristics, including the following categories and subcategories:

> **Order:** The highest category in the soil classification systems. The criteria distinguishing the soil orders are based on the kinds of horizons and their degree of development. Accordingly, twelve orders are recognized: Alfisols, andisols, aridisols, entisols, gelisols, histosols, inceptisols, mollisols, oxisols, spodosols, ultisols, vertisols.

> **Suborder:** This category specifies the ranges of soil moisture and temperature regimes, and the kind and composition of the horizons.

> **Great group:** Groups of soils that have similar horizons as well as similar moisture and temperature regimes.

> **Subgroup:** Soils that have distinguishing characteristics of their own within the great group.

> **Family:** Soils that have particular physical and mineralogical properties that affect plant growth.

> **Series:** Subdivisions of soil families that are most similar in their main profile characteristics.

Spatial variability of properties and processes: The non-uniform distribution of specifiable attributes of an area, whether random or systematic, and the range of their variation expressed by means of statistical criteria.

Specific heat: The change in heat content of a unit mass of a substance or body per unit change in temperature. Of the various substances in the soil, water exhibits the highest specific heat.

Specific surface: The total surface area of particles per unit mass of the particles or per unit volume of the soil, usually expressed as square meters per gram or per cubic centimeter.

Splash erosion: The action of raindrops on bare soil, in which the impacting drops detach soil particles and splash them in all directions but with a net tendency toward the downslope direction. This lashing action initiates the process of water erosion.

Stefan-Boltzmann law: The total energy flux emitted by a radiating body, integrated over all wavelengths, is proportional to the fourth power of the absolute temperature of the body's surface.

Stokes law: An equation relating the velocity of a spherical particle settling in an aqueous suspension, under the influence of gravity, to the particle radius and density, as well as to the viscosity and density of the fluid.

Strain: The ratio of the deformation of a body that is subject to stress (compressive, tensile, or shear) to the body's original dimensions.

Strength: The capacity of a body to withstand stresses without experiencing failure (whether by rupture, fragmentation, collapse, or flow). In quantitative terms, it is the maximal stress a given body can bear without undergoing failure, or the minimal stress that will cause the body to fail.

Stress: A force per unit area acting on a body and tending to deform it by compression, tension, or shear.

Structural stability: The ability of a soil's aggregated structure to resist the forces tending to cause its disruption, such as slaking by water or grinding and compaction by traffic.

Structure of soil: The arrangement and organization of the particles in the soil (i.e., the internal configuration of the soil matrix, or fabric), whether single-grained (unattached particles variously stacked) or associated in more or less stable aggregates. As such, the structure of the soil determines the geometric configuration of the pores and the permeability of the soil to fluids.

Surface-water excess: The excess of rainfall intensity over the infiltration rate. That excess tends to accumulate in pockets or depressions over the surface, or – if the surface is smooth and sloping – to run off the surface in the form of overland flow.

Symbiosis: Two or more species living in close association, to mutual benefit. An example is a class of bacteria that are attached to the roots of leguminous plants and are able to supply the plants with nitrogen fixed from the atmosphere, while obtaining sugars in return.

Tensiometer: A device to measure the matric potential of soil moisture in situ, consisting of a porous cup filled with water, with a manometer to monitor the pressure of the water in the cup as it equilibrates with soil moisture. As soil moisture diminishes, its matric tension increases, hence it draws water from the cup, which in turn registers of subatmospheric pressure, called tension (or suction).

Tension plate apparatus: A device for equilibrating an initially saturated soil sample with a known matric tension value, applicable to the tension range of 0 to 1 bar.

Terracing: The practice of building horizontal barriers on the countour, aimed at checking surface runoff and preventing it from accelerating downslope. Thus, a long slope is divided into sections, which help to retain rainwater and minimize erosion.

Texture of soil: The range of particle sizes in a soil, expressed in terms of the proportions by mass of the fractions known as sand (the coarsest particles), silt (intermediate-size particles), and clay (the finest particles).

Thermal conductivity: The flux of heat (i.e., the amount of heat conducted across a unit cross-sectional area in unit time) per unit temperature gradient.

Thermal regime of the soil profile: The variation of soil temperature with time for various depths in the profile.

Thermocouple psychrometer: A device to measure relative humidity of the air phase in the soil.

Tillage: Mechanical manipulation of the soil's upper layer by means of implements designed to slice, pulverize, loosen, invert, and/or mix the soil. The aims of tillage are to prepare a seedbed, eradicate weeds, enhance infiltration and aeration, and/or shape the soil surface. However, tillage operations consume much energy (fossil fuel for the operation of machinery), disrupt soil structure, and hasten the decomposition of organic matter. Recent trends are to minimize or eliminate unnecessary tillage.

Time-domain reflectometry: A technique to measure soil wetness, consisting of parallel metal rods inserted into the soil. When a step pulse of electricity is sent through the rods, the received signal is related to the dielectric properties of the medium. Since the dielectric constant of water is about 81 while that of soil solids is 4 to 8 and that of air is about 1, the reading depends mainly on the fractional volume of water present.

Topsoil: The upper layer of the soil, generally the layer richest in organic matter and nutrients, but also the most vulnerable to erosion.

Tortuosity: The ratio of the average "roundabout" path of flow in a porous medium to the apparent, or straight, flow path.

Transpiration: Evaporation from plants, principally through the open stomates in the leaves. When plants experience water stress, they tend to close their stomates, thereby restricting transpiration. By so doing, however, they also curtail the process of photosynthesis, i.e., the absorption of atmospheric carbon dioxide by the stomates.

Turbulent flow: The internally disordered flow of a fluid such that "packets" of the fluid swirl about in eddies, and more energy is dissipated in internal friction than in the case of laminar flow.

Two-phase flow: The simultaneous flow in the soil of two fluids, ether immiscible (e.g., air and water, or petroleum and water) or miscible (e.g., saline and fresh water).

Universal soil loss equation: An empirical equation that considers the rate of erosion as a combined product of rainfall erosivity, soil erodibility, slope steepness or length, surface cover, and soil management.

Vadose zone: The unsaturated soil and the underlying porous strata that ovelie the permanent water table. It is also called the "unsaturated zone" or the "aerated zone," in contrast with the saturated zone that lies underneath the water table.

Vapor pressure: The partial pressure of water vapor in the atmosphere. It depends on the pressure and temperature of the atmosphere, as well as on the state of water in a water-containing body (such as soil) at equilibrium with the atmosphere.

Vertisols: Soil rich in expansive clay (e.g., smectite). When subject to alternating cycles of wetting and drying, as in a semiarid region, such soils tend to heave and then to settle and form wide, deep cracks, as well as slanted sheer planes extending deep into the soil profile.

Viscosity: The resistance of a fluid to shear (i.e., to forces causing adjacent layers of the fluid to slide over each other). The resistance is proportional to the velocity of the shearing. As such, viscosity can be visualized as the fluid's internal friction.

Void ratio: Fractional volume of voids (pores) per volume of soil solids.

Water application efficiency: The net amount of water added to the root zone as a fraction of the total amount applied to the field. Surface irrigation methods such as flooding or furrow irrigation may result in runoff and/or percolation beyond the root zone, thus reducing the application efficiency. Sprinkling irrigation may involve additional losses due to wind drift. Microirrigation methods, if well managed, offer the potential for relatively high application efficiency.

Water balance of a field: An equation expressing the law of mass conservation, summing all quantities of water added to, subtracted from, and stored within a given volume of soil (e.g., the root zone of a cropped field or of a natural plant habitat, or even an entire watershed) during a specified period of time.

Water conveyance efficiency: The volume of water arriving at the field as a fraction of the volume taken from some source (say, a reservoir). The difference between those volumes represents the loss of water incurred in conveyance and distribution, such as seepage and evaporation from open (often unlined) canals or leakage from pipes.

Water potential in the soil: A measure of the energy state of water in the soil relative to that of pure water at atmospheric pressure and a standard elevation. At each point, water potential depends on temperature, hydrostatic pressue (being "negative" in unsaturated soil, hence called tension or suction), solute concentration, and relative elevation.

Water stability of aggregates: The ability of soil aggregates to avoid disintegration (or "slaking") when saturated by, or submerged in, water.

Water table: The top surface of a body of groundwater, that surface being at atmospheric pressure. All groundwater beneath the water table is at a pressure greater than atmospheric, whereas soil moisture above the water table is at a pressure smaller than atmospheric (i.e., a subpressure called tension or suction). A distinction should be made between the main (regional) water table and the possible occurrence of a "perched" water table that rests (generally temporarily) on an relatively impervious stratum above the main water table.

Water use efficiency: The measure of crop produced (total vegetative growth or net marketable product) per unit amount of water applied or per unit amount of water consumed by the crop.

Waterlogged soil: A soil that is nearly saturated with water much of the time, such that anaerobic conditions prevail. In extreme cases of prolonged waterlogging, anaerobiosis occurs, the roots of mesophytes suffer, and chemical reduction processes occur (including denitrification, methanogenesis, and the reduction of iron and manganese oxides).

Watershed (catchment): A secton of the land surface that drains its surface-water excess (i.e., the excess of rainfall over infiltration) to a specified point along a stream.

Watershed divide: The line delineating the upper edge of a watershed and separating between it and adjacent watersheds.

Wetness: The water content per unit mass of soil solids (mass wetness) or per unit bulk volume of the soil (volume wetness).

Wettability: The tendency of a normal soil to sorb water (in contrast with the tendency of some soils to repel water, called "hydrophobicity"). Water applied to a wettable soil typically forms an acute contact angle with the solid surfaces, and is drawn into the soil's pores.

Windbreaks: Strips of trees or fences built of slats or brush, arranged perpendicularly to the direction of the prevailing wind, aimed at reducing its speed so as to prevent wind erosion and desiccation, as well as mechanical damage to crops and roads by drifting sand and snow. Vegetated windbreaks are also called shelterbelts.

Xerophytes: Plants adapted to growing in arid regions, able to survive prolonged droughts.

Zonal soil: A mature residual soil formed under local climatic conditions and reflecting the cumulative and lasting influences of its formative conditions. Such a soil is assumed to be at equilibrium with the climate as the principal soil-forming factor.

BIBLIOGRAPHY

Adams, R. McC. (1981). Heartland of Cities. University of Chicago Press, Chicago.

Addiscott, R.M., Whitmore, A.P., and Powlson, D.S. (1991). Farming, Fertilizers, and the Nitrate Problem. CAB International, Wallingford, UK.

Agren, C.I., and Bosatta, E. (1997). Theoretical Ecosystem Ecology: Understanding Element Cycles. Cambridge University Press, Cambridge, UK.

Alexander, M. (1977). Introduction to Soil Microbiology, 2nd edition, John Wiley, NY.

Alexander, M. (1994). Biodegradation and Bioremediation, Academic Press, San Diego, CA.

Alexandros, N., Editor (1995). World Agriculture Toward 2010. FAO, Rome.

Aloway, B.J., Editor. (1990). Heavy Metals in Soils. Blackie, Glasgow, UK.

Artzy, M., and Hillel, D. (1988). A Defense of the Theory of Progressive Soil Salinization in Ancient Southern Mesopotamia. Geoarchaeology 3:235–238.

Assink, J.W. and van den Brink, W.J., Editors (1986). Contaminated Soils. Martinus Nijhof, Dordrecht.

Baker, R.S., Gee, G.W., and Rosenzweig, C., Editors. (1994). Soil and Water Science: Key to Understanding Our Global Environment. Soil Science Society of America, Madison, WI.

Baluska, F., Ciamporova, M., and Barlow, P.W., Editors. (1995). Structure and Function of Roots. Kluwer Academic, Dordrecht, Netherlands.

Barber, S.A. (1995). Soil Nutrient Bioavailability: A Mechanistic Approach. Wiley, New York.

Bell, M. and Boardman, J., Editors. (1992). Past and Present Soil Erosion: Archaeological and Geographical Perspectives. Oxbow Books, Oxford.

Binns, T., Editor (1995). People and Environment in Africa. Wiley, Chichester.

Bird, R.B., Stewart, W., and Lightfoot, E. (2002). Transport Phenomena. Wiley, New York, NY.

Blevins, R.L., and Fry, W.W. (1992). Conservation Tillage: An Ecological Approach to Soil Management. Advances in Agronomy 51:33–78.

Boeker, E., and van Gondelle, R. (1995). Environmental Physics. Wiley, Chichester, UK.

Boggs, S. (1995). Principles of Sedimentation and Stratigraphy. Prentice Hall, Englewood Cliffs, NJ.

Borowski, O. (1987). Agriculture in Iron Age Israel. Eisenbrauns, Winona Lake, IN.

Boserup, E. (1981). Population and Technological Change: A Study of Long-Term Trends. Chicago University Press, Chicago.

Bottema, S., Entjes-Nieborg, G., and Van Zeist, W., Editors (1990). Man's Role in the Shaping of the Eastern Mediterranean Landscape. Balkema, Rotterdam.

Bouwman, A.F., Editor. (1990). Soils and the Greenhouse Effect: Climate Change and Ecosystems. Wiley, Chichester, UK.

Bowman, A.K. and Kogan, E., Editors (1999). Agriculture in Egypt from Pharaonic to Modern Times. Oxford University Press, Oxford.

Bradford, J.M., and Peterson, G.A. (2000). Conservation Tillage. In: Sumner, M.E., Editor. Handbook of Soil Science. CRC Press, Boca Raton, Fl.

Brady, N.C., and Weil, R.R. (1996). The Nature and Properties of Soils. Prentice Hall, Upper Saddle River, NJ.

Bras, R.L. (1990). Hydrology: An Introduction to Hydrologic Science. Addison Wesley, Reading, MA.

Bresler, E., McNeal, B.L., and Carter, D.L. (1982). Saline and Sodic Soils: Principles-Dynamics-Modeling. Springer, Berlin, Germany.

Briggs, D., and Courtney, F. (1985). Agriculture and Environment: The Physical Geography of Agricultural Systems. Longman, Harlow, UK.

Brown, J.R., Editor (1987). Soil Testing, Sampling, Correlation, Calibration, and Interpretation. Soil Science Society of America, Madison, Wisconsin.

Brussaard, L., and Ferrera-Cerrato, R., Editors. (1997). Soil Ecology in Sustainable Agricultural Systems. Lewis Publishers, Boca Raton, Fl.

Buol, S.W., Hole, F.D., and McCracken, R.J. (1989). Soil Genesis and Classification. Iowa State University Press, Ames, IA.

Buol, S.W., Hole, F.D., McCracken, R.J., and Southard, R.J. (1997). Soil Genesis and Classification. Iowa State University Press, Ames, Iowa.

Buresh, R.J., Sanchez, P.A., and Calhoun, F., Editors (1997). Replenishing Soil Fertility in Africa. Soil Science Society of America, Madison, Wisconsin.

Burke, W., Gabriels, D., and Bouma, J. (1986). Soil Structure Assessment. Balkema, Roterdam, Netherlands.

Butzer, K.W. (1976). Early Hydraulic Civilization in Egypt. University of Chicago Press, Chicago.

Campbell, G.S., and Norman, J.N. (1998). An Introduction to Environmental Biophysics. Springer, New York, NY.

CAST (1980). Organic and Conventional Farming Compared. Council for Agricultural Science and Technology. Ames, IA.

CAST (1982). Soil Erosion: Its Agricultural, Environmental, and Socioeconomic Implications. Council for Agricultural Science and Technology. Ames, IA.

Chancellor, W.J., Editor. (1994). Advances in Soil Dynamics, Volume 1. Monograph 12, American Society of Agricultural Engineers, St. Joseph, MI.

Clay, J. (2004). World Agriculture and the Environment. Island Press, Washington, DC.

Cohen, J.E. (1995). How Many People Can the Earth Support? Norton, New York.

Coleman, D.C., Crossley, D.A., and Hendrix, P.F. (2004). Fundamentals of Soil Ecology. Academic Press, San Diego, CA.

Collins, W.W., and Qualset, C.O. (1999). Biodiversity in Agroecosystems. CRC, Boca Raton, FL.

Cowan, C.W. and Watson, P.J., Editors (1992).The Origin of Agriculture: An International Perspective. Smithsonian Institute Press, Washington, D.C.

Coyne, M. (1999). Soil Microbiology: An Exploratory Approach. Thomson Delmar, Clifton Park, NY.

Craul, P.J. (1999). Urban Soils: Applications and Practices. Wiley, New York, NY.

Dale, T., and Carter, V.G. (1955). Topsoil and Civilization. University of Oklahoma Press, Norman, OK.

Dane, J.H. and Topp, G.C., Editors. (2002). Methods of Soil Analysis, Part 4: Physical Methods. Soil Science Society of America, Madison, WI.

De Haan, F.A.M. and Visser-Reineveld, M.I., Editors (1996). Soil Pollution and Soil Protection. International Training Centre, State Agricultural University, Wageningen, Netherlands.

De Ploey, J., Editor. (1983). Rainfall Simulation, Runoff, and Soil Erosion. Catena, Cremlingen, Germany.

De Wit, C.T., and Goudriaan, J. (1978). Simulation of Ecological Processes. PUDOC, Wageningen, Netherlands.

Dexter, A.R. (1987). Mechanics of Root Growth. Plant Soil 98: 303–312.

Doran, J.W., Coleman, D.C., Bezdicek D.F., and Stewart B.A., (1994). Defining soil quality for a sustainable environment. Soil Science Society of America Special Publication Number 35, ASA, Madison Wis.

Doran, J.W., Sarrantonio, M. and Liebig, M.lS. (1996). Soil health and sustainability. Advances in Agronomy 56:1–54.

Dubbin, W. (2001). Soils. The Natural History Museum, London, UK.

Epstein, E. (1977). The Role of Roots in the Chemical Economy of Life on Earth. Bioscience 27:783–787.

Evans, L.T. (1998). Feeding the Ten Billion: Plants and Population Growth. Cambridge University Press, Cambridge.

Faleknmark, M. (1997). Meeting water requirements of an expanding world population. Philosophical Transactions of the Royal Society of London B352:901–906.

FAO (1990). Soil Map of the World. U.N. Food and Agriculture Organization, Rome, Italy.

FAO (1995). Irrigation in Africa in Figures. Water Report No. 7, FAO, Rome.

Farr, E., and Henderson, W.C. (1986). Land Drainage. Longman, London, UK.

Faulkner, E.H. (1943). Plowman's Folly. Grosser and Dunlap, New York.

Fischer, G. and Heilig, G.K. (1997). Population momentum and the demand on land and water resources. Philosophical Transactions of the Royal Society of London B352:869–889.

Fley, J.A. (1995). An Equilibrium Model of the Terrestrial Carbon Budget. Tellus 371:52–54.

Frissel, M.J., Editor (1978). Cycling of Mineral Nutrients in Agricultural Systems. Elsevier, Amsterdam.

Fussell, G.E. (1965). Farming Techniques from Prehistory to Modern Times. Pergamon Press, Oxford.

Gardner, W.R. (1996). A New Era for Irrigation. National Academy Press, Washington, DC.

Gleick, P.H., Editor (1993). Water in Crisis: A guide to the World's Fresh Water Resources. Oxford University Press, New York.

Gobat, J-M., Aragno, M., and Matthey, W. (2004). The Living Soil: Fundamentals of Soil Science and Soil Biology. Science Publishers, Enfield, NH.

Goedert, W.J. (1983). Management of the cerrado soils of Brazil: a review. Journal of Soil Science 34:405–428.

Goudie, A. (1995). The Changing Earth. Blackwell, Oxford.

Goudie, A. (2000). The Human Impact on the Natural Environment. MIT Press, Cambridge, MA.

Goulding, K.W.T. (2000). Nitrate leaching from arable and horticultural land. Soil Use and Management 16:145–151.

Greenland, D.J. and Szabolcs, I., Editors (1994). Soil Resilience and Sustainable Land Use. CAB International, Wallingford, UK.

Gregorich, E.G. and Carter, M.R., Editors (1997). Soil Quality for Crop Production and Ecosystem Health. Developments in Soil Science No. 25, Elsevier, Amsterdam.

Gregorich, E.G., Turchenek, L.W., Carter, M.R., and Angers, D.A. (2002). Soil and Environmental Science Dictionary. CRC Press, Boca Raton, FL.

Grigg, D.B. (1974). The Agricultural Systems of the World: An Evolutionary Approach. Cambridge University Press, Cambridge.

Harley, J.I. and Smith, S.E. (1983). Mycorrhizal Symbiosis. Academic Press, London.

Harpstead, M.I., Sauer, T.J., and Bennett, W.F. (2001). Soil Science Simplified. Iowa State University Press, Ames, IA.

Harris, D.R., Editor (1996). The Origins and Spread of Agriculture and Patoralism in Eurasia. University College Press, London.

Hartemink, A.E., Editor (2006). The Future of Soil Science. International Union of Soil Science, Wageningen, Netherlands.

Hatenrath, S. (1991). Climate Dynamics of the Tropics. Kluwer, Dordrecht.

Haub, C. (1995). How many people have ever lived on Earth? Population Today, Feb.

Hayward, M.D., Bosemark, N.O. and Romangosa, I., Editors (1003). Plant Breeding Principles and Prospects. Chapman and Hall, London.

Helms, D. (1984). Walter Lowdermilk's journey: Forester to land conservationist. Environmental Review *:112–45.

Hemond, H.F., and Fechner-Levy, E.J. (2000). Chemical Fate and Transport in the Environment. Academic Press, San Diego, CA.

Henry, D.O. (1989). From Foraging to Agriculture: The Levant at the End of the Ice Age. University of Pennsylvania Press, Philadelphia.

Henson, R. (2006). The Rough Guide to Climate Change. Penguin, London, UK.

Hill, R.L. (1990). Long-term Conventional and No-till Effects on Selected Soil Physical Properties. Soil Sci. Soc. Am. J. 54:161–166.

Hillel, D. (1977). Computer Simulation of Soil Water Dynamics. International Development Research Centre, Ottawa, Canada.

Hillel, D. (1991). Out of the Earth: Civilization and the Life of the Soil. University of California Press, Berkeley, CA.

Hillel, D. (1995). Rivers of Eden: The Struggle for Water and the Quest for Peace in the Middle East. Oxford University Press, New York.

Hillel, D. (1997). Small-Scale Irrigation for Arid Zones. UN Food and Agriculture Organization, Rome, Italy.

Hillel, D. (2000). Salinity Management for Sustainable Irrigation. The World Bank, Washington, DC.

Hillel, D. (2004). Introduction to Environmental Soil Physics. Academic Press, San Diego, CA.

Hillel, D. (2006). The Natural History of the Bible: An Environmental Exploration of the Hebrew Scriptures. Columbia University Press, New York, NY.

Hillel, D., and Rosenzweig, C. (2002). Desertification and Its Relation to Climate Variability and Change. Advances in Agronomy 77:1–39. Academic Press, San Diego, CA.

Hillel, D., and Rosenzweig, C. (2005). The Role of Biodiversity in Agronomy. Advances in Agronomy 88:1–35. Academic Press, San Diego, CA.

Hillel, D., Editor (1982–1987). Advances in Irrigation, Volumes I, II, III, IV. Academic Press, San Diego, CA.

Hillel, D., Editor-in-Chief (2005). Encyclopedia of Soils in the Environment. Elsevier, Oxford, UK.

Hooke, R.L. (2000). On the history of humans as geomorphic agents. Geology 28:843–6.

Houghton, J.T., Jenkins, G.J., and Ephraums, J.J. (1998). Intergovernmental Panel on Climate Change. Cambridge University Press, Cambridge, UK.

Hughes, J.D. (1975). Ecology in Ancient Civilizations. University of New Mexico Press, Albuquerque.

Hyams, E. (1976). Soil and Civilization. Harper and Row, New York.

IPCC (Intergovernmental Panel on Climate Change), (1995). Climate Change 1994: Scenarios. Cambridge University Press, Cambridge, UK.

Jackson, I.J. (1989). Climate, Water, and Agriculture in the Tropics. Longman, Harlow.

Jackson, W. (2002). Natural systems agriculture: a truly radical alternative. Agriculture, Ecosystems, and Environment 88:111–7.

Jacobsen, Th. And Adams, R. McC. (1958). Salt and silt in ancient Mesopotamian agriculture. Science 128:1251–1258.

Jenkinson, D.S., Adams, D.E., and Wild, A. (1991). Model estimates of CO_2 emissions from soil in response to global warming. Nature 351:304–306.

Jenny, H. (1941). Factors of Soil Formation: A System of Quantitative Pedology. McGraw-Hill, New York.

Joffe, J.S. (1949). Pedology. Pedology Publications, New Brunswick, NJ.

Kemp, D.D. (1994). Global Environmental Issues: A Climatological Approach. Routledge, London, UK.

Kemp, R.J. (1989). Ancient Egypt: Anatomy of a Civilization. Routledge, London.

Killham, K. (1994). Soil Ecology. Cambridge University Press, Cambridge.

Kirkby, M.H. and Morgan, R.P.C., Editors (1980). Soil Erosion. Wiley, Chichester.

Kramer, P.J., and Boyer, J.S. (1995). Water Relations of Plants and Soils. Academic Press, San Diego, CA.

Kutilek, M., and Nielsen, D.R. (1994). Soil Hydrology. Catena, Cremlingen, Germany.

Laegreid, M., Bockman, O.C., and Kaarstad, O. (1999). Agriculture, Fertilizers, and the Environment. CAB International, Wallingford, UK.

Lal, R. (1987). Tropical Ecology and Physical Edaphology. John Wiley, Chicester, UK.

Lal, R. (1989). Conservation tillage for sustainable agriculture: tropics versus temperate environments. Advances in Agronomy 42:85–197.

Lal, R. (2004). Soil carbon sequestration impacts on global climate change and food security. Science 304:1623–7.

Lal, R., Editor (1993). World Soil Erosion and Conservation. Cambridge University Press, Cambridge.

Lal, R., Kimble, J.M., Follett, R.F., and Stewart, B.A., Editors. (1998). Management of Carbon Sequestration in Soil. CRC Press, Boca Raton, Fl.

Lamb, H.H. (1995). Climate, History, and the Modern World. Routledge, London.

Lavelle, P., and Spain, A.V. (2005). Soil Ecology. Springer, Dordrecht, Netherlands.

Le Houerou, H.N. (1996). Climate change, drought, and desertification. Journal of Arid Environemnts 34:133–85.

Lowdermilk, W.C. (1953). Conquest of the Land Through 7,000 Years. Bulletin 99, Soil Conservation Service, U.S. Department of Agriculture, Washington, DC.

Malthus, T.R. (1798). An Essay on the Principle of Population. Republished (1986), William Pickering, London.

Marschner, H. (1986). Mineral Nutrition of Higher Plants. Academic Press, San Diego, CA.

McBride, M.B. (1994). Environmental Chemistry of Soils. Oxford University Press, Oxford.

McDonald, A., and Kay, D. (1988). Water Resources: Issues and Strategies. Longman, Harlow.

McGregor, G.R. and Nieuwolt, G. (1998). Tropical Climatology: An Introduction to the Climates of Low Latitudes. Wiley, Chichester.

McNeill, J.R., and Winimarter, V., Editors. Soils and Societies: Perspectives from Environmental History. The White Horse Press, Isle of Harris, UK.

Mellars, P. (2004). Neanderthals and the modern human colonization of Europe. Nature 432:461–5.

Miller, R.D. (1980). Freezing Phenomena in Soils. In: Hillel, D., Applications of Soil Physics. Academic Press, New York, NY.

Milliman, J.D., Yun-Shan, Q., Mei-E, R., and Saito, Y. (1987). Man's influence on the erosion and transport of sediment by Asian rivers: The Yellow River (Huanghe) as an example. Journal of Geology 95:751–62.

Monteith, J. (1975). Principles of Environmental Physics. Arnold, London.

Monteith, J.L. (1980). The Development and Extension of Penman's Evaporation Formula. In: Hillel, D., Applications of Soil Physics. Academic Press, New York, NY.

Moore, A.M.T. and Hillman, G.C. (1992). The Pleistocene to Holocene transition and human economy in Southwest Asia: The impact of the Younger Dryas. American Antiquity 57:482–94.

Morgan, R.P.C. (1995). Soil Erosion and Conservation. Longman, Harlow.

Mosier, A.R., Syers, K., and Freney, J.R. (2004). Agricutlure and the Nitrogen Cycle. Island Press, Washington, DC.

Muller, C. (1999). Modeling Soil-Biosphere Interactions. CABI Publishing, Wallingford, UK.

Nair, P.K.R. (1993). An Introduction to Agroforestry. Kluwer Academic, Dordrecht.

Nalven, G.F., Editor. (1997). The Environment: Air, Water, and Soil. American Institute of Chemical Engineering. New York, NY.

National Research Council (1993a). Sustainable Agriculture and the Environment in the Humid Tropics. National Academy Press, Washington, D.C.

National Research Council (1993b). Soil and Water Quality: An Agenda for Agriculture. National Academy Press, Washington, D.C.

Nearing, M.A., Pruski, E.G., and O'Neal, M.R. (2004). Expected climate change impacts on soil erosion rates: A review. Journal of Soil and Water Conservation 59:43–50.

Nebel, B.J. (1987). Environmental Science: The Way the World Works. Prentice-Hall, Englewood Cliffs, NJ.

Nye, P.H. and Greenland, D.J. (1960). The Soil Under Shifting Cultivation. Commonwealth Bureau of Soils, Harpenden, UK.

O'Neill, B.C., MacKellar, F.L., and Lutz, W. (2001). Population and Climate Change. Cambridge University Press, Cambridge.

Oldeman, L.R., Hakkeling, R.T.A., and Sombroek, W.G. (1991). World Map of the Status of Human-Induced Soil Degradation: Am Explanatory Note. International Soil Reference and Information Centre, Wageningen, Netherlands.

Olson, R., Francis, C., and Kaffka, S., Editors. (1995). Exploring the Role of Diversity in Sustainable Agriculture. American Society of Agronomy, Madison, WI.

Partridge, M. (1973). Farm Tools Through the Ages. Oxprey, Reading.

Paul, P.A., and Clark, F.E. (1996). Soil Microbiology and Biochemistry, 2nd edition, Academic Press, San Diego, CA.

Peixoto, J.P., and Oort, J.H. (1992). Physics of Climate. American Physics Institute, New York, NY.

Pepper, I.L., Gerba, C.P., and Brusseau, M.L., Editors. (2006). Environmental and Pollution Science. Elsevier, Amsterdam, Netherlands.

Pereira, H.C. (1973). Land Use and Water Resources in Temperate and Tropical Climates. Cambridge University Press, Cambridge.

Pessarakli, M. (1996). Handbook of Plant and Crop Physiology. Marcel Dekker, New York, NY.

Philip, J.R. (1995). Desperately Seeking Darcy in Dijon. Soil Sci. Soc. Am. J. 59:319–324.

Pierzynski, G.M., Sims, J.T., and Vance, G.F. (2000). Soils and Environmental Quality. CRC Press, Boca Raton, Florida.

Pimentel, D. (1997). Pest Management in Agriculture. In: Techniques for Reducing Pesticide Use. John Wiley, New York, NY.

Pimentel, D., Editor (1993). World Soil Erosion and Conservation. Cambridge University Press, Cambridge.

Ponting, C. (1992). A Green History of the World: The Environment and the Collapse of Great Civilizations. St. Martin's Press, New York.

Powlson, D.S., Smith, P., and Smith, J.U., Editors (1996). Evolution of Soil Organic Matter Models Using Existing Long-Term Databases. Springer-Verlag, Berlin.

Pringle, H. (1998). Neolithic agriculture: The slow birth of agriculture. Science 282:1446.

Rechcigl, J.E., Editor (1995). Soil Amendments: Impacts on Biotic Systems. CRC Press, Boca Raton, FL.

Redman, C.L. (1978). The Rise of Civilization. Freeman, San Francisco.

Rengel, Z., Editor (1999). Mineral Nutrition of Crops: Fundamental Mechanisms and Implications. Haworth Press, New York.

Reuss, J.O. and Johnson, D.W. (1986). Acid Deposition and the Acidification of Soils and Waters. Springer-Verlag, New York.

Richards, R. (1987). The Microbiology of Terrestrial Ecosystems, Longman Scientific & Technical.

Rindos, D. (1984). The Origins of Agriculture: An Evolutionary Perspective. Academic Press, New York.

Roberts, N. (1998). The Holocene: An Environmental History. Blackwell, Oxford.

Rosenzweig, C., and Hillel, D. (1998). Climate Change and the Global Harvest: Potential Impacts of the Greenhouse Effect on Agriculture. Oxford University Press, New York, NY.

Rosenzweig, C., and Hillel, D. (2007). Climate Variability and the Global Harvest: Impacts of El Nino and Other Oscillations on Agroecosystems. Oxford University Press, New York, NY.

Rothberg, R.I. and Rabb, T.K., Editors (1985). Hunger and History. Cambridge University Press, Cambridge.

Sanchez, P.A. (1976). Properties and Management of Soils in the Tropics. Wiley-Interscience, New York.

Sanchez, P.A., Buresh, R.J., and Leakey, R.R.B. (1997). Trees, soils, and food security. Philosophical Transactions of the Royal Society of London B352:949–961.

Scharopenseel, H.W., Schomaker, M., and Ayoub, A., Editors. (1990). Soils on a Warmer Earth. Elsevier, Amsterdam, Netherlands.

Schlesinger, W.H. (1991). Biogeochemistry: An Analysis of Global Change. Academic Press, San Diego, CA.

Schultz, J. (2005). The Ecozones of the World. Springer-Verlag, Heidelberg, Germany.

Singer, M.J., and Munns, D.N. (1987). Soils: An Introduction. Macmillan, New York, NY.

Skipper, H.D. and Turco, R.F., Editors (1995). Bioremediation: Science and Applications. Soil Science Society of America, Madison, Wisconsin.

Smartt, J. and Simmonds, N.W., Editors (1995). Evolution of Crop Plants. Longman, Harlow.

Smil, V. (1997). Cycles of Life: Civilization and the Biosphere. Scientific American Library, New York, NY.

Smil, V. (2000). Feeding the World: A Challenge for the Twenty-First Century. MIT Press, Cambridge, MA.

Smil, V. (2001). Enriching the Earth: Fritz Haber, Carl Bosch, and the Transformation of World Food Production. MIT Press, Cambridge, MA.

Smith, S.E. and Read, D.J. (1997). Mycorrhizal Symbiosis. Academic Press, San Diego.

Soane, B.D. and Ouwerkerk, C., Editors (1994). Soil Compaction in World Agriculture. Elsevier, Amsterdam.

Soil and Water Conservation Society (2000). Soil Biology Primer. Ankey, OH.

Soil Science Society of America (1996). Glossary of Soil Science Terms. SSSA, Madison, WI.

Soil Survey Division Staff (1993). Soil Survey Manual. Handbook 18, US Department of Agriculture, Washington, DC.

Soil Survey Staff (1999). Soil taxonomy: A basis system of soil classification for making and interpreting soil surveys. Natural Resources Conservation Service, USDA, Agricultural Handbook No. 436, Washington, DC.

Soil Survey Staff (2003). Keys to Soil Taxonomy. Ninth Edition. U.S. Government Printing Office, Washington DC.

Sparks, D.L. (2003). Environmental Soil Chemistry. Academic Press, San Diego, CA.

Sposito, G. (1989). The Chemistry of Soils. Oxford University Press. New York, NY.

Sprent, J.I. and Sprent, P. (1990). Nitrogen Fixing Organisms: Pure and Applied Aspects. Chapman and Hall, London.

Stanhill, G. (1986). Irrigation in arid lands. Philosophical Transactions of the royal Society of London A316:261–273.

Stephens, D.B. (1995). Vadose Zone Hydrology. Lewis Publishers, Boca Raton, FL.

Stevenson, F.J. and Cole M.A. (1999). Cycles of Soil Carbon, Nitrogen, Phosphorus, Sulfur, and Micronutrients. Wiley, New York.

Stewart, B.A. and Nielsen, D.R., Editors (1990). Irrigation of Agricultural Crops. American Society of Agronomy, Madison, Wisconsin.

Sumner, M.E., Editor-in-Chief. (2000). Handbook of Soil Science. CRC Press, Boca Raton, Fl.

Syivitski, J.P., Vorosmarty, C.J., Kertner, A.J., and Green, P. (2005). Impacts of humans on the flux of terrestrial sediment to the global coastal ocean. Science 308:376–80.

Sylvia, D.M. et al., 1998, Principles and Applications of Soil Microbiology, Prentice Hall.

Szaboles, I. (1979). Salt-Affected Soils. UNESCO, Paris.

Tan, K.H. (2003). Humic Matter in Soil and the Environment: Principles and Controversies. Marcel Dekker, New York, NY.

Tanji, K.K., Editor. (1990). Agricultural Salinity Assessment and Management. American Society of Civil Engineering. New York, NY.

Tate, R.I. (2000). Soil Microbiology, 2nd edition, John Wiley.

Tiesen, H., Cuevas, E., and Chacon, P. (1994). The role fo soil organic matter in sustaining soil fertility. Nature 371:783–5.

Tinker, P.B. (1997). The environmental implications of intensified land use in developing countries. Philosophical Transactions of the Royal Society of London B352:1023–1033.

Topp, E., and Pattey, E. (1997). Soils as Sources and Sinks for Atmospheric Methane. Can. J. Soil Sci. 77:167–178.

Trudgill, S.T. (1988). Soil and Vegetation Systems. Clarendon Press, Oxford, UK.

U.N. Food and Agriculture Organization (FAO) (2000). World Soil Map. Rome, Italy.

Unger, P.W. (1990). Conservation Tillage Systems. In: Singh, R.P., Editor. Dryland Agriculture: Strategies for Sustainability. Advances in Soil Science, 13. Springer, New York, NY.

United Nations (1999). World Population Prospects. United Nations, New York.

USDA, National Resources Conservation Service (2006). Global Soil Regions. Washington, DC.

USDA-NRCS - Soil Biology URL accessed on 2006-04-11

van Elsas et al., 1997, Modern Soil Microbiology, Marcel Dekker.

Vitousek, P.M., Mooney, H.A., LUbchenko, J., and Melillo, J.M. (1997). Human domination of Earth's Ecosystems. Science 277:494–9.

Wallace, A., and Terry, R.E., Editors. (1997). Handbook of Soil Conditioners: Substances that Enhance the Physical Properties of Soil. Marcel Dekker, New York, NY.

Warkentin, B.P., Editor. (2006). Footprints in the Soil: People and Ideas in Soil History. Elsevier, Amsterdam, Netherlands.

Warrick, A.W. (2003). Soil Water Dynamics. Oxford University Press, New York, NY.

Wassmann, R., and Vlek, P.L.G., Editors. (2004). Tropical Agriculture in Transition – Opportunities for Mitigating Greenhouse Gas Emissions. Kluwer, Dordrecht, Netherlands.

Wayne, R.P. (1991). Chemistry of the Atmosphere. Clarendon Press, Oxford, UK.

Webster, R. (1997). Soil resources and their assessment. Philosophical Transactions of the Royal Society of London B352:963–973.

Wild, A. (1994). Soils and the Environment. Cambridge University Press, Cambridge, UK.

Wild, A., Editor (1991). Russell's Soil Conditions and Plant Growth. Longman, Harlow.

Williams, M. (2003). Deforesting the Earth: From Prehistory to Global Crisis. University of Chicago Press, Chicago.

Wischmeier, W.H. and Smith, D.D. (1978). Predicting Rainfall Erosion Losses. USDA Agricultural Handbook 537, United States Department of Agriculture, Washington, D.C.

Wolman, M.G. and Fourner, F.G.A., Editors (1987). Land Transformation in Agriculture. Wiley, Chichester.

Wood, M. (1995). Environmental Soil Biology, 2nd edition. Blackie, Glasgow, UK.

Woodwell, G.M. (1990). The Earth in Transition: Patterns and Processes of Biotic Impoverishment. Cambridge University Press, Cambridge, UK.

Woomer, P.L. and Swift, M.J., Editors (1994). The Biological Management of Tropical Soil Fertility. Wiley, Chichester.

Young, A. (1997). Agroforestry for Soil Management. CAB International, Wallingford, UK.

Young, A. (1998). Land Resources Now and for the Future. Cambridge University Press, Cambridge.

Young, L.Y., and Cerniglia, C.E., Editors. (1995). Microbial Transformations of Toxic Organic Chemicals. Wiley, New York, NY.

INDEX

Note: Page numbers followed by *b*, *f*, *n*, *t* indicates boxes, figures, notes and tables respectively.

FIGURE SOURCES
AND CREDITS

Fig. 1.1 from: Hillel, D. (1991). Out of the Earth: Civilization and the Life of the Soil. University of California Press, Berkeley.

Fig. 2.1 from: Hillel, D. (1994). Rivers of Eden: The Struggle for Water and the Quest for Peace in the Middle East. Oxford University Press, New York.

Fig. 2.2 from: Hillel, D. (2006). The Natural History of the Bible: An Environmental Exploration of the Hebrew Scriptures. Columbia University Press, New York.

Fig. 3.1 from: Dubbin, W. (2001). Soils. The Natural History Museum, London.

Fig. 3.2 from: Hillel, D. (1991). Out of the Earth: Civilization and the Life of the Soil. University of California Press, Berkeley.

Fig. 3.3 from: Coleman and Crossley (1996). Fundamentals of Soil Ecology. Academic Press, San Diego.

Fig. 3.4 from: Hillel, D. (2004). Introduction to Environmental Soil Physics. Academic Press/Elsevier, San Diego.

Fig. 3.5 from: Hillel, D. (2004). Introduction to Environmental Soil Physics. Academic Press/Elsevier, San Diego.

Fig. 4.1 from: Schultz, J. (2005). Ecozones of the World: The Ecological Divisions of the Geosphere. Springer Verlag, Berlin.

Fig. 4.2 from: Schultz, J. (2005). Ecozones of the World: The Ecological Divisions of the Geosphere. Springer Verlag, Berlin.

Fig. 4.3 from: Schultz, J. (2005). Ecozones of the World: The Ecological Divisions of the Geosphere. Springer Verlag, Berlin.

Fig. 4.4 from: Nebel, B.J. (1987). Environmental Science: The Way the World Works. Prentice-Hall, Englewood Cliffs, New Jersey.

Fig. 4.5 from: Natural Resources Conservation Service (2006), U.S. Department of Agriculture, Washington, DC.

Fig. 4.6 from: Natural Resources Conservation Service (2006), U.S. Department of Agriculture, Washington, DC.

Fig. 4.7 from: Nebel, B.J. (1987). Environmental Science: The Way the World Works. Prentice-Hall, Englewood Cliffs, New Jersey.

Fig. 4.8 from: Natural Resources Conservation Service (2006), U.S. Department of Agriculture, Washington, DC.

Fig. 5.1 from: Hillel, D. (2004). Introduction to Environmental Soil Physics. Academic Press/Elsevier, San Diego.

Fig. 5.2 from: Hillel, D. (2004). Introduction to Environmental Soil Physics. Academic Press/Elsevier, San Diego.

Fig. 5.3 from: Hillel, D. (2004). Introduction to Environmental Soil Physics. Academic Press/Elsevier, San Diego.

Fig. 5.4 from: Hillel, D. (2004). Introduction to Environmental Soil Physics. Academic Press/Elsevier, San Diego.

Fig. 5.5 from: Hillel, D. (2004). Introduction to Environmental Soil Physics. Academic Press/Elsevier, San Diego.

Fig. 5.6 from: Hillel, D. (2004). Introduction to Environmental Soil Physics. Academic Press/Elsevier, San Diego.

Fig. 5.7 from: Hillel, D. (2004). Introduction to Environmental Soil Physics. Academic Press/Elsevier, San Diego.

Fig. 5.8 from: Hillel, D. (2004). Introduction to Environmental Soil Physics. Academic Press/Elsevier, San Diego.

Fig. 5.9 from: Hillel, D. (2004). Introduction to Environmental Soil Physics. Academic Press/Elsevier, San Diego.

Fig. 5.10 from: Hillel, D. (2004). Introduction to Environmental Soil Physics. Academic Press/Elsevier, San Diego.

Fig. 5.11 from: Hillel, D. (2004). Introduction to Environmental Soil Physics. Academic Press/Elsevier, San Diego.

Fig. 5.12 from: Hillel, D. (2004). Introduction to Environmental Soil Physics. Academic Press/Elsevier, San Diego.

Fig. 5.13 from: Dubbin, W. (2001). Soils. The Natural History Museum, London.

Fig. 5.14 from: Hillel, D. (2004). Introduction to Environmental Soil Physics. Academic Press/Elsevier, San Diego.

Fig. 6.1 from: Hillel, D. (2004). Introduction to Environmental Soil Physics. Academic Press/Elsevier, San Diego.

Fig. 6.2 from: Hillel, D. (1977). Computer Simulation of Soil Water Dynamics. International Development Research Centre, Ottawa, Canada.

Fig. 6.3 from: Hillel, D. (2004). Introduction to Environmental Soil Physics. Academic Press/Elsevier, San Diego.

Fig. 6.4 from: Hillel, D. (2004). Introduction to Environmental Soil Physics. Academic Press/Elsevier, San Diego.

Fig. 6.5 from: Hillel, D. (2004). Introduction to Environmental Soil Physics. Academic Press/Elsevier, San Diego.

Fig. 6.6 from: Hillel, D. (2004). Introduction to Environmental Soil Physics. Academic Press/Elsevier, San Diego.

Fig. 7.1 from: Hillel, D. (2004). Introduction to Environmental Soil Physics. Academic Press/Elsevier, San Diego.

Fig. 7.2 from: Hillel, D. (2004). Introduction to Environmental Soil Physics. Academic Press/Elsevier, San Diego.

Fig. 7.3 from: Hillel, D. (2004). Introduction to Environmental Soil Physics. Academic Press/Elsevier, San Diego.

Fig. 7.4 from: Hillel, D. (2004). Introduction to Environmental Soil Physics. Academic Press/Elsevier, San Diego.

Fig. 7.5 from: Hillel, D. (2004). Introduction to Environmental Soil Physics. Academic Press/Elsevier, San Diego.

Fig. 8.1 from: Hillel, D. (2004). Introduction to Environmental Soil Physics. Academic Press/Elsevier, San Diego.

Fig. 8.2 from: Hillel, D. (2004). Introduction to Environmental Soil Physics. Academic Press/Elsevier, San Diego.

Fig. 8.3 from: Hillel, D. (2004). Introduction to Environmental Soil Physics. Academic Press/Elsevier, San Diego.

Fig. 8.4 from: Hillel, D. (2004). Introduction to Environmental Soil Physics. Academic Press/Elsevier, San Diego.

Fig. 8.5 from: Hillel, D. (2004). Introduction to Environmental Soil Physics. Academic Press/Elsevier, San Diego.

Fig. 8.6 from: Hillel, D. (2004). Introduction to Environmental Soil Physics. Academic Press/Elsevier, San Diego.

Fig. 8.7 from: Hillel, D. (2004). Introduction to Environmental Soil Physics. Academic Press/Elsevier, San Diego.

Fig. 8.8 from: Hillel, D. (2004). Introduction to Environmental Soil Physics. Academic Press/Elsevier, San Diego.

Fig. 8.9 from: Hillel, D. (2004). Introduction to Environmental Soil Physics. Academic Press/Elsevier, San Diego.

Fig. 8.10 from: Hillel, D. (2004). Introduction to Environmental Soil Physics. Academic Press/Elsevier, San Diego.

Fig. 8.11 from: Hillel, D. (2004). Introduction to Environmental Soil Physics. Academic Press/Elsevier, San Diego.

Fig. 8.12 from: Hillel, D. (2004). Introduction to Environmental Soil Physics. Academic Press/Elsevier, San Diego.

Fig. 8.13 from: Hillel, D. (2004). Introduction to Environmental Soil Physics. Academic Press/Elsevier, San Diego.

Fig. 9.1 from: Hillel, D. (2004). Introduction to Environmental Soil Physics. Academic Press/Elsevier, San Diego.

Fig. 9.2 from: Hillel, D. (2004). Introduction to Environmental Soil Physics. Academic Press/Elsevier, San Diego.

Fig. 9.3 from: Hillel, D. (1997). Small-Scale Irrigation for Arid Zones: Principles and Options. Food and Agriculture of the U.N., Rome.

Fig. 9.4 from: Rosenzweig, C. and Hillel, D. (1998). Climate Change and the Global Harvest: Potential Impacts of the Greenhouse Effect on Agriculture. Oxford University Press, New York.

Fig. 9.5 from: Hillel, D. (2004). Introduction to Environmental Soil Physics. Academic Press/Elsevier, San Diego.

Fig. 9.6 from: Gates, D. (1980). Biophysical Ecology. Springer Verlag, New York.

Fig. 9.7 from: Hillel, D. (2004). Introduction to Environmental Soil Physics. Academic Press/Elsevier, San Diego.

Fig. 9.8 from: Hillel, D. (2004). Introduction to Environmental Soil Physics. Academic Press/Elsevier, San Diego.

Fig. 10.1 from: Schultz, J. (2005). Ecozones of the World: The Ecological Divisions of the Geosphere. Springer Verlag, Berlin.

Fig. 10.2 from: Dubbin, W. (2001). Soils. The Natural History Museum, London.

Fig. 10.3 from: Hillel, D. (2004). Introduction to Environmental Soil Physics. Academic Press/Elsevier, San Diego.

Fig. 10.4 from: Hillel, D. (2000). Salinity Management for Sustainable Irrigation: Integrating Science, Environment, and Economics. The World Bank, Washington, DC.

Fig. 11.1 from: Schultz, J. (2005). Ecozones of the World: The Ecological Divisions of the Geosphere. Springer Verlag, Berlin.

Fig. 11.2 from: Harpstead, M.I., Sauer, T.J., and Bennett, W.F. (2001). Soil Science Simplified. Iowa State University Press, Ames, IA.

Fig. 11.3 from: Anon. (1977). Soils: An Outline of Their Properties and Management. CSIRO, Melbourne, Australia.

Fig. 12.1 from: Gobat, J-M., Aragno, M., and Matthey, W. (2004). The Living Soil: Fundamentals of Soil Science and Soil Biology. Science Publishers, Enfield, NH.

Fig. 12.2 from: Nebel, B.J. (1987). Environmental Science: The Way the World Works. Prentice-Hall, Englewood Cliffs, New Jersey.

Fig. 12.3 from: Gobat, J-M., Aragno, M., and Matthey, W. (2004). The Living Soil: Fundamentals of Soil Science and Soil Biology. Science Publishers, Enfield, NH.

Fig. 12.4 from: Schultz, J. (2005). Ecozones of the World: The Ecological Divisions of the Geosphere. Springer Verlag, Berlin.

Fig. 13.1 from: Hillel, D. (1998). Environmental Soil Physics. Academic Press, San Diego.

Fig. 13.2 from: Hillel, D. (1998). Environmental Soil Physics. Academic Press, San Diego.

Fig. 13.3 from: Hillel, D. (1998). Environmental Soil Physics. Academic Press, San Diego.

Fig. 13.4 from: Hillel, D. (1998). Environmental Soil Physics. Academic Press, San Diego.

Fig. 13.5 from: Hillel, D. (2004). Introduction to Environmental Soil Physics. Academic Press/Elsevier, San Diego.

Fig. 13.6 from: Hillel, D. (2004). Introduction to Environmental Soil Physics. Academic Press/Elsevier, San Diego.

Fig. 13.7 from: Hillel, D. (2004). Introduction to Environmental Soil Physics. Academic Press/Elsevier, San Diego.

Fig. 14.1 from: Dubbin, W. (2001). Soils. The Natural History Museum, London.

Fig. 14.2 from: Hillel, D. (1998). Environmental Soil Physics. Academic Press, San Diego.

Fig. 14.3 from: Hillel, D. (1998). Environmental Soil Physics. Academic Press, San Diego.

Fig. 14.4 from: FAO (1987). Protect and Produce. UN Food and Agriculture Organization, Rome.

Fig. 14.5 from: Hillel, D. (2004). Introduction to Environmental Soil Physics. Academic Press/Elsevier, San Diego.

Fig. 14.7 from: Schultz, J. (2005). Ecozones of the World: The Ecological Divisions of the Geosphere. Springer Verlag, Berlin.

Fig. 14.8 from: Nebel, B.J. (1987). Environmental Science: The Way the World Works. Prentice-Hall, Englewood Cliffs, New Jersey.

Fig. 14.9 from: Hillel, D. (2004). Introduction to Environmental Soil Physics. Academic Press/Elsevier, San Diego.

Fig. 15.1 from: Hillel, D. (2004). Introduction to Environmental Soil Physics. Academic Press/Elsevier, San Diego.

Fig. 15.2 from: Nebel, B.J. (1987). Environmental Science: The Way the World Works. Prentice-Hall, Englewood Cliffs, New Jersey.

Fig. 15.3 from: Nebel, B.J. (1987). Environmental Science: The Way the World Works. Prentice-Hall, Englewood Cliffs, New Jersey.

Fig. 15.4 from: Rubin, H. (2005). Groundwater Pollution. In: Hillel, D., Editor-in-Chief, Encyclopedia of Soils in the Environment, Vol. 3:271–281. Elsevier, Oxford.

Fig. 15.5 from: Nebel, B.J. (1987). Environmental Science: The Way the World Works. Prentice-Hall, Englewood Cliffs, New Jersey.

Fig. A.1 from: Dubbin, W. (2001). Soils. The Natural History Museum, London.

Fig. A.2 from: Rosenzweig, C. and Hillel, D. (1998). Climate Change and the Global Harvest: Potential Impacts of the Greenhouse Effect on Agriculture. Oxford University Press, New York.

Fig. A.3 from: Gobat, J-M., Aragno, M., and Matthey, W. (2004). The Living Soil: Fundamentals of Soil Science and Soil Biology. Science Publishers, Enfield, NH.

Fig. A.4 from: Dubbin, W. (2001). Soils. The Natural History Museum, London.